AF779247

Dynamic Insights: Unveiling the Biomechanics of Sport Through Motion Analysis

Dynamic Insights: Unveiling the Biomechanics of Sport Through Motion Analysis

Guest Editors

Ukadike Chris Ugbolue
Sang-Hoon Yeo

MDPI

Basel • Beijing • Wuhan • Barcelona • Belgrade • Novi Sad • Cluj • Manchester

Guest Editors

Ukadike Chris Ugbolue
School of Health and Life Sciences
University of the West of
Scotland
Scotland
UK

Sang-Hoon Yeo
School of Sport
Exercise and Rehabilitation Sciences
University of Birmingham
Birmingham
UK

Editorial Office
MDPI AG
Grosspeteranlage 5
4052 Basel, Switzerland

This is a reprint of the Special Issue, published open access by the journal *Bioengineering* (ISSN 2306-5354), freely accessible at: https://www.mdpi.com/journal/bioengineering/special_issues/L8Z3D8RV61.

For citation purposes, cite each article independently as indicated on the article page online and as indicated below:

Lastname, A.A.; Lastname, B.B. Article Title. *Journal Name* **Year**, *Volume Number*, Page Range.

ISBN 978-3-7258-7743-0 (Hbk)
ISBN 978-3-7258-7744-7 (PDF)
https://doi.org/10.3390/books978-3-7258-7744-7

© 2026 by the authors. Articles in this reprint are Open Access and distributed under the Creative Commons Attribution (CC BY) license. The reprint as a whole is distributed by MDPI under the terms and conditions of the Creative Commons Attribution-NonCommercial-NoDerivs (CC BY-NC-ND) license (https://creativecommons.org/licenses/by-nc-nd/4.0/).

Contents

About the Editors

Ukadike Chris Ugbolue

Ukadike Chris Ugbolue is a Lecturer in Applied Biomechanics at the University of the West of Scotland (UWS) and a member of the Sport Performance and Active Living research community. He also holds an honorary position as Associate Professor at Ningbo University, China. His research focuses on human movement analysis, sports biomechanics, and the application of emerging technologies to enhance athletic performance, rehabilitation, and health. His expertise includes motion capture, electromyography, wearable sensing technologies, and biomechanical modelling to examine movement patterns in sport, clinical, and functional settings.

Dr Ugbolue's work spans interdisciplinary areas including sports performance analysis, injury prevention, neuromuscular function, and rehabilitation technologies. A key aspect of his research involves translating biomechanical insights into practical applications that support athletes, clinicians, and rehabilitation specialists. His research has contributed to understanding movement mechanics in sports such as golf, boxing, and running, alongside broader investigations into fatigue, neuromuscular control, and musculoskeletal health.

In addition to his academic research, Dr Ugbolue is actively engaged in innovation and translational research. He is currently developing assistive technologies, including a rehabilitation glove designed to support stroke recovery and improve upper-limb function. This work integrates biomechanics, wearable technology, and clinical rehabilitation principles to develop practical solutions for patient care.

Dr Ugbolue has authored and co-authored numerous peer-reviewed publications in biomechanics, sports science, and bioengineering, and collaborates with international researchers and industry partners. Through his research, teaching, and editorial activities, he contributes to advancing the understanding of human movement and technology-driven approaches to sport performance and rehabilitation.

Sang-Hoon Yeo

Sang-Hoon Yeo is a Lecturer in Biomechanics and Motor Control in the School of Sport, Exercise and Rehabilitation Sciences at the University of Birmingham and a member of the Centre for Human Brain Health. His research focuses on understanding how the brain, muscles, and skeletal system interact to control human movement, with the goal of improving diagnosis, rehabilitation, and performance in health and disease.

Dr Yeo has an interdisciplinary background in mechanical engineering, computer science, and biomechanics. He completed his BSc and MSc in Mechanical and Aerospace Engineering at Seoul National University, South Korea, before undertaking a PhD in Computer Science at the University of British Columbia, Canada, where he specialised in neuromusculoskeletal modelling and simulation. Prior to joining the University of Birmingham, he worked as a research associate in the Sensorimotor Learning Group at the University of Cambridge.

His research integrates biomechanics, robotics, virtual reality, and computational modelling to investigate sensorimotor learning, eye-hand coordination, and muscle mechanics. A central focus of his work is the development of computational and experimental approaches to understand how humans acquire, retain, and adapt motor skills. He also leads projects exploring new muscle mechanics models, immersive technologies for concussion assessment, and affordable home-based neuro-motor rehabilitation systems for individuals with neurological conditions such as stroke or Parkinson's disease.

Dr Yeo has authored numerous peer-reviewed publications across biomechanics, neuroscience, and computational modelling. Through interdisciplinary collaborations spanning engineering, sport science, and health research, his work contributes to advancing knowledge of human movement and developing innovative technologies for movement assessment and rehabilitation.

Preface

This Special Issue explores how motion analysis and emerging technologies advance the understanding of sport biomechanics, performance, and rehabilitation. It brings together research on movement analysis, fatigue, and technology-driven assessment, offering insights for researchers, clinicians, and practitioners seeking to improve athlete development and injury prevention.

Ukadike Chris Ugbolue and Sang-Hoon Yeo
Guest Editors

bioengineering

MDPI

Article

The Role of Lower Limb Kinetics in Boxing Punches and the Impact of Fatigue on Biomechanical Performance

Charles Stewart [1,2], Ross Cornett [1,2], Julien S. Baker [1,3], Yaodong Gu [1], Frédéric Dutheil [4] and Ukadike Chris Ugbolue [1,2,*]

1 Faculty of Sports Science, Ningbo University, Ningbo 315211, China; b00191475@studentmail.uws.ac.uk (C.S.); juliensbakerw1@gmail.com (J.S.B.)
2 School of Health and Life Sciences, Sport and Physical Activity Research Institute (SPARI), University of the West of Scotland, South Lanarkshire, Scotland G72 0LH, UK
3 Centre for Health and Exercise Science Research, Department of Sport, Physical Education and Health, Hong Kong Baptist University, Kowloon Tong, Hong Kong
4 National Centre for Scientific Research (CNRS), LaPSCo, Physiological and Psychosocial Stress, University Hospital of Clermont-Ferrand, CHU Clermont-Ferrand, Preventive and Occupational Medicine, WittyFit, Université Clermont Auvergne, 63000 Clermont-Ferrand, France
* Correspondence: u.ugbolue@uws.ac.uk; Tel.: +44-(0)1698-283100 (ext. 8284)

Abstract

Purpose: This study investigated the contribution of lower limb kinetics to punch performance in amateur boxing and examined the effects of fatigue on biomechanical efficiency. **Methods:** Ten male amateur boxers performed six punch types (jab, cross, left hook, right hook, left uppercut, right uppercut) under non-fatigued and post-fatigue conditions. Ground reaction force (GRF) and rate of force development (RFD) were measured using dual force plates, while punch outputs were assessed via a boxing force sensor. Fatigue was induced using a 9.5 min lower-body circuit. **Results:** Pre-fatigue, the cross punch generated the highest outputs for punch force (1475.42 N), GRF (947.54 N), and RFD (3973.38 N/s). Post-fatigue, punch force declined significantly across all punches (–4.26%, $p = 0.027$), with the greatest reductions in the cross and left hook. RFD responses were variable, with compensatory increases observed in some punches. Intra-individual analysis revealed greater fatigue-induced declines in the weakest punches (–9.84%, $p = 0.001$) compared with the strongest (–4.63%, $p = 0.027$). **Conclusions:** Lower limb force generation, particularly rear-leg drive, is critical to punch effectiveness and fatigue resilience. Conditioning programs should prioritise lower limb endurance while addressing performance variability across punch types.

Keywords: boxing; ground reaction force; rate of force development; fatigue; biomechanics; kinetic chain

1. Introduction

Despite longstanding reports of physical trauma and its acute and chronic neurological and neuropsychological effects on the brain, boxing remains popular [1,2]. Boxing is a multidimensional combat sport that requires the integration of technical skill, strength, coordination and endurance to deliver high-impact strikes while maintaining defensive capability [3,4]. Such investigations across different experience levels are important, as Zazryn and colleagues [5] reported that injuries occur more frequently in amateur boxing than in professional boxing. Central to punching performance is the ability to generate force rapidly and transfer it efficiently through the kinetic chain-from the lower limbs to the upper extremities-culminating at the punch point. Ground reaction force (GRF) and rate of force development (RFD) are critical components of this process, particularly in rotational and straight-line strikes where force production originates from the lower body and trunk [6,7].

Research suggests that boxers who optimise the sequential activation of proximal-to-distal segments achieve greater punch impact [8]. However, the biomechanical literature remains disproportionately focused on upper-body kinematics, despite evidence that lower limb force contributes significantly to punching force and stability [7,9]. GRF generated by the rear leg, in particular, is a key determinant of punch power during the cross and hook [9]. RFD further reflects the explosive capacity of the legs, and its relationship with punching efficiency has been highlighted in multiple striking sports [3,10].

Fatigue is another crucial factor that may disrupt this biomechanical sequence. Lower body fatigue has been shown to reduce postural control, impair kinetic transfer and alter strike mechanics [6]. Yet, the degree to which punch performance deteriorates under fatigue-particularly in amateur boxers-remains underexplored. Some evidence indicates that boxers may maintain output through compensatory upper body adjustments [9], while others suggest significant reductions in both GRF and punch force after fatigue [7].

Few studies have examined fatigue-induced biomechanical changes across multiple punch types. The extent to which different strikes (e.g., jab vs. uppercut) are affected by fatigue may reflect the varying kinetic demands of each punch. Additionally, no previous research has compared intra-individual variability in punch performance-specifically, how a boxer's weakest and strongest punches respond to fatigue. Understanding this variability could offer practical insight for conditioning interventions aimed at improving consistency and reducing performance drop-off.

This study aimed to examine the role of lower limb kinetics (GRF and RFD) in six common boxing punches-jab, cross, left hook, right hook, left uppercut and right uppercut-and to determine the effect of lower body fatigue on punch output. A novel element of this investigation was the intra-subject comparison between each boxer's strongest and weakest punches. Findings from this study are intended to inform training strategies that enhance lower limb endurance and minimise intra-individual variability in strike performance.

2. Materials and Methods

2.1. Study Design

A within-subject repeated-measures design was employed to investigate the contribution of lower limb kinetics in punching performance and the effect of fatigue on this relationship. Each participant completed two testing conditions: a non-fatigued punch assessment and a post-fatigue punch assessment conducted immediately after a standardised lower limb fatigue circuit. This design controlled for inter-individual variability and enabled direct comparison of biomechanical outputs under fatigued and non-fatigued states. All testing was conducted under standardised conditions in a university biomechanics laboratory (Figure 1).

Figure 1. Experimental setup for punch force assessment. (**A**) Front view of a participant in the standardized stance position wearing boxing gloves and displaying the lower limb Plug-in Gait (PiG) retro-reflective marker set., (**B**) Back view of participant stance with boxing glove with (PiG) marker set, (**C**) Display of the mounted boxing pad attached to a fixed frame for stability, and (**D**) Digital display used to record punch force values.

2.2. Participants

Ten right-hand-dominant male amateur boxers (mean age: 25.2 ± 3.8 years; height: 178.6 ± 6.5 cm; body mass: 78.4 ± 7.9 kg) were recruited from local boxing gyms and university sport science networks. All participants had a minimum of two years' boxing experience and were free from injury at the time of data collection. Ethical approval was obtained from the University of the West of Scotland Ethics Committee (Approval Number: 15871) and written informed consent was secured from all participants.

2.3. Kinetic Data Collection

Kinetic data were collected using a Kistler force plate system (Kistler Instruments Ltd., Hampshire, UK), sampling at 1000 Hz. Participants adopted their natural boxing stance, with each foot positioned on a force plate throughout the punch execution. Ground reaction force (GRF) data were recorded across three axes: vertical (Fz), anterior-posterior (Fy) and medial-lateral (Fx). Primary analyses focused on the Fz and Fy components due to their relevance to propulsion and force transmission [11].

Rate of force development (RFD) was calculated using the standard formula:

$$\text{RFD} = (\Delta\text{Force (N)})/(\Delta\text{Time (s)})$$

This provided a measure of the explosive capacity of force production during punching actions. Kinetic output was segmented into four distinct phases: Base, Load, Drive and Impact. All testing was performed at the same time of day to control for the effects of diurnal variation.

2.4. Punch Force Output

The examination of force distribution and centre of pressure displacement is a common approach for analysing motion, loading, and load distribution in biomechanical research. Unlike gait analysis, the progression of force during boxing punches represents a relatively new area of investigation. Understanding centre of pressure displacement and the distribution of forces across the fist during a punch is essential for examining both the effects of impact on the hand's biological structures and the technical biomechanical factors underpinning punching performance [12]. Punch force was assessed using the Hoteam Boxing Force Sensor Target Machine [13], which outputs impact values in kilograms. These values were converted to Newtons. While not equivalent in accuracy to laboratory-grade motion capture or force-measuring targets, the device offered consistent output during pilot testing and was deemed suitable for applied boxing assessments. Punch velocity was not measured as the focus of this study focused on kinetic-rather than kinematic-variables.

2.5. Fatigue Protocol

To simulate the lower limb fatigue experienced during boxing competition, participants completed two rounds of a structured bodyweight circuit. Participants performed the exercises at a self-selected pace. Each round lasted 4.5 min and included the following exercises:

Static Wall Sit (30 s).
Standing Calf Raises (30 s).
Bodyweight Squats (30 s).
Box Step-Ups (30 s).
Walking Lunges (30 s).

Each exercise was followed by a 30 s static wall sit. A 30 s active recovery period separated the two rounds, resulting in a total circuit duration of 9 min and 30 s. Participants continued the protocol until they either reached a rating of perceived exertion (RPE) of $\geq$18 on the Borg 6–20 scale or were physically unable to continue. Exercise selection was based on previous combat sports fatigue models [14].

Before and after the fatigue protocol, participants performed two sets of three maximal-effort punches for each of the following types: jab, cross, hook and uppercut. Technique was maintained across efforts, and each punch was executed with maximal intent.

2.6. Punch Selection and Categorisation

For each punch type, the highest and lowest force efforts were identified for each participant to examine intra-individual variability. These values were extracted from six repetitions (three per pre/post condition) per punch type. The strongest efforts were retained for further GRF and RFD analysis, while the weakest efforts were used to compare punch output changes under fatigue.

2.7. Data Processing and Statistical Analysis

GRF and RFD values were computed across each of the four defined punch phases. Punch force values were manually recorded from the Hoteam sensor and averaged across the three repetitions per punch type. Descriptive statistics (Mean ± SD) were calculated for all dependent variables. Strongest and weakest punch forces (Mean ± SD) were also reported.

Paired-sample t-tests were used to assess pre- vs. post-fatigue differences in GRF, RFD and punch output. The difference between the two means was examined. Cohen's D (d) was used to express the size of the effect of the pre- vs. post-fatigue differences with effect size thresholds set as 0.2 = small, 0.5 = medium and 0.8 = large [15]. Additional paired-sample t-tests were conducted to compare fatigue-related changes between the strongest and weakest punches. Statistical significance was set at $p < 0.05$.

3. Results

3.1. Ground Reaction Forces Across All Punches (Pre-Fatigue)

Rear and front leg vertical ground reaction forces (GRFs) were recorded across movement phases during the non-fatigued condition. Rear leg GRF consistently peaked during the drive phase and declined sharply by impact across all punch types. For the jab, rear GRF reached 1243.41 ± 494.64 N during drive and dropped to 44.53 ± 33.07 N at impact. The front leg GRF increased from 362.44 ± 59.78 N at base to 418.47 ± 383.74 N at impact.

In the cross, rear leg GRF peaked at 1135.53 ± 363.79 N during drive and reduced to 90.88 ± 52.70 N at impact. The front leg GRF ranged from 442.22 ± 119.88 N at base to 472.97 ± 280.90 N at impact. For the left and right hooks, peak rear drive GRFs were 1040.68 ± 389.90 N and 1103.82 ± 478.79 N, with impact values of 97.49 ± 52.41 N and 83.37 ± 36.32 N, respectively. Corresponding front impact GRFs were 453.68 ± 288.70 N and 432.17 ± 266.22 N.

Rear drive GRF for the left and right uppercuts was 963.83 ± 312.14 N and 1028.22 ± 337.17 N, respectively, dropping to 64.32 ± 41.79 N and 85.15 ± 47.62 N at impact. Front leg GRF at impact was 408.24 ± 197.34 N (left uppercut) and 425.09 ± 230.45 N (right uppercut).

3.2. Punch Output Across All Punches (Pre-Fatigue)

Pre-fatigue punch force output varied across punch types. The cross produced the highest mean force (1475.42 ± 261.40 N), followed by the left hook (1374.38 ± 203.13 N), right hook (1326.31 ± 188.12 N), right uppercut (1241.95 ± 167.52 N), jab (1126.25 ± 291.67 N) and left uppercut (1007.49 ± 120.06 N).

Weakest punch efforts were also identified for intra-individual comparison. The weakest jab averaged 900.56 ± 218.52 N, while the weakest cross measured 1196.82 ± 278.35 N. Weak left and right hooks produced 1024.16 ± 189.73 N and 999.64 ± 186.39 N, respectively. The lowest forces were seen in the left uppercut (736.73 ± 143.69 N) and right uppercut (973.15 ± 180.52 N). The mean punch force across all pre-fatigue efforts was 1275.30 ± 251.43 N.

3.3. Rate of Force Development Across All Punches (Pre-Fatigue)

The left hook produced the highest mean RFD (2676.91 ± 1387.60 N/s), followed by the jab (2547.45 ± 1172.35 N/s) and the cross (2517.17 ± 1541.06 N/s). The right hook measured 2326.94 ± 1279.71 N/s. RFD was lowest for the left uppercut (1781.53 ± 1185.00 N/s) and right uppercut (1561.46 ± 953.19 N/s). Straight punches and hooks exhibited greater explosive force characteristics than uppercuts.

3.4. Ground Reaction Forces Across All Punches (Post-Fatigue)

Post-fatigue GRF patterns followed similar trends, with notable reductions at impact. For the jab, rear GRF during drive was 1370.62 ± 403.26 N and dropped to 578.69 ± 349.74 N at impact. Front leg GRF ranged from 374.81 ± 104.83 N at base to 338.55 ± 316.66 N at impact.

In the cross, rear drive GRF increased to 1324.59 ± 371.12 N, and rear impact dropped to 644.93 ± 301.28 N. Front leg values ranged from 345.17 ± 95.30 N to 500.68 ± 305.33 N. The left and right hooks recorded rear drive GRFs of 1125.18 ± 374.77 N and 1173.55 ± 449.99 N, with rear impact at 596.24 ± 283.10 N and 548.62 ± 288.70 N, respectively. Front leg impact values were 461.47 ± 278.20 N (left hook) and 455.61 ± 267.53 N (right hook).

The left and right uppercuts produced rear drive GRFs of 1061.29 ± 312.88 N and 1059.45 ± 348.36 N, with rear impact values of 499.90 N and 498.53 N. Front impact GRFs were 428.08 ± 198.47 N (left) and 443.91 ± 225.64 N (right).

3.5. Punch Output Across All Punches (Post-Fatigue)

Punch force declined across all types as the effect size ranged from extremely small (i.e., very close to zero/practically negligible) to medium. The jab decreased to 1133.05 ± 238.56 N (–0.6%, p = 0.111, d = –0.026), the cross to 1369.48 ± 190.01 N (–7.18%, p = 0.029, d = 0.464), and the left hook to 1261.57 ± 202.80 N (–8.21%, p = 0.0049, d = 0.556). The right hook declined to 1295.90 ± 166.64 N (–2.29%, p = 0.126, d = 0.171). The left uppercut fell to 968.25 ± 125.80 N (–3.89%, p = 0.292, d = 0.319), and the right uppercut to 1174.26 ± 164.42 N (–5.45%, p = 0.013, d = 0.408).

In the weakest punch group, jab output fell to 811.29 ± 353.49 N (–9.91%, p = 0.326, d = 0.304), cross to 1039.86 ± 220.77 N (–13.11%, p = 0.067, d = 0.625), left hook to 938.82 ± 126.31 N (–8.33%, p = 0.083, d = 0.530), and right hook to 956.48 ± 135.08 N (–4.32%, p = 0.192, d = 0.265). The left uppercut dropped to 634.71 ± 126.82 N (–13.85%, p = 0.056, d = 0.753), and the right uppercut to 876.03 ± 134.67 N (–9.98%, p = 0.016, d = 0.610). The difference between the pre and post groups ranged from small to large effects.

3.6. Rate of Force Development Across All Punches (Post-Fatigue)

Post-fatigue RFD showed variable responses. The jab increased to 3420.59 ± 1178.30 N/s (+34.28%, p = 0.0116). The cross showed a significant decline to 2029.56 ± 1594.39 N/s (–24.03%, p = 0.027). The left and right hooks declined to 2327.65 ± 2339.72 N/s (–13.05%, p = 0.664) and 2158.99 ± 1798.33 N/s (–7.22%, p = 0.620), respectively. The left uppercut increased to 2307.95 ± 1425.87 N/s (+29.55%, p = 0.081), while the right uppercut showed the largest increase, rising to 2324.30 ± 1372.39 N/s (+48.85%, p = 0.021).

3.7. Comparative Effects of Fatigue on Strongest and Weakest Punches

The strongest punches declined from 1258.63 ± 170.79 N to 1200.42 ± 141.82 N (–4.63%, $p = 0.027$). The weakest punches showed a greater drop, from 971.84 ± 151.32 N to 876.20 ± 141.20 N (–9.84%, $p = 0.001$) (Figure 2). Among the weakest punches, the left uppercut (–13.85%) and cross (–13.11%) exhibited the largest reductions.

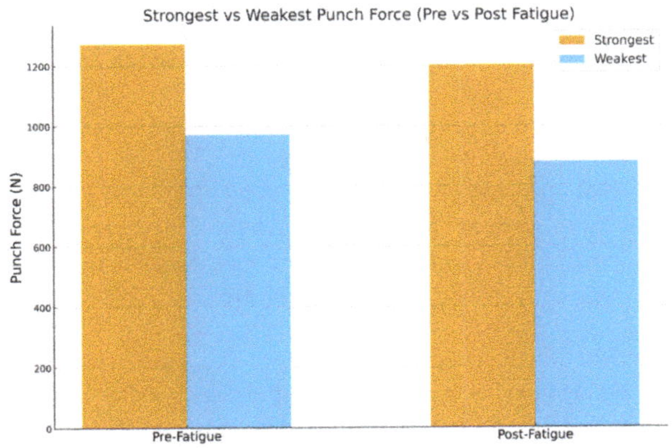

Figure 2. Histogram representation of the Comparative Effects of Fatigue on the Strongest and Weakest Punches.

3.8. Calibration and Validation of the Hoteam Boxing Force Sensor

The Hoteam Boxing Force Sensor was calibrated and validated prior to data collection. Static calibration was performed using traceable weights (0 kg–150 kg) and a Kistler Force Plate to generate a calibration curve across the full measurement range; the calibration was linear across the operational range ($R^2 = 0.98$). Dynamic validation consisted of synchronous recordings of 10 controlled (20 kg) strikes using a mechanical striker performed by a laboratory technician onto a calibrated Kistler force platform. The results produced a coefficient of variation of 4.57%.

4. Discussion

This study investigated the role of lower limb kinetics, specifically ground reaction force (GRF) and rate of force development (RFD), in boxing punch performance and examined how these variables respond to fatigue. Six punch types were assessed using force plates and a punch force sensor in both pre and post fatigue conditions. The results confirmed that GRF peaked during the drive phase, particularly through the rear leg and that the cross and left hook produced the highest punch outputs. Following fatigue, both punch force and GRF generally declined, although RFD responses varied [7,9].

A notable strength of this study was the intra-individual comparison between each participant's strongest and weakest punches across all six strike types. These comparisons revealed significant variability in performance, with the weakest punches consistently producing markedly lower outputs. For example, the left hook exhibited the greatest decline post-fatigue (−8.21%) and the largest difference between strongest and weakest trials (−22.8%). These findings reinforce the importance of efficient kinetic chain sequencing and suggest that this efficiency is vulnerable to disruption under fatigue [6,16], supporting earlier research on rear-leg contributions and whole-body coordination in effective striking [9].

4.1. Lower Limb Kinetics and Punch Performance

Peak GRF consistently occurred during the drive phase, with the rear leg contributing most significantly to vertical (Fz) and anterior-posterior (Fy) force production. Pre-fatigue, the cross delivered the highest punch force (1475.42 N), GRF (947.54 N), and RFD (3973.38 N/s), followed closely by the left hook (1374.38 N, 868.53 N, and 3797.01 N/s, respectively). These findings align with Stanley et al. [7] and Liu et al. [10], who reported elevated GRF values in experienced boxers during rotational punches.

The jab recorded the lowest punch force (1126.25 N) and GRF (791.76 N) but produced high RFD (2919.91 N/s), indicating a compensatory reliance on rapid force production rather than total force. This supports McGill et al. [3], who highlighted the importance of upper-body neuromuscular coordination in short-range strikes.

Uppercuts, particularly the left uppercut, yielded the lowest GRF (700.28 N) and RFD (2559.28 N/s), potentially due to their more vertical trajectory and limited leg drive involvement [6]. These variations in kinetic output highlight the distinct mechanical demands of each punch type and advocate for punch-specific training approaches [10,17,18].

4.2. Fatigue Effects on Kinetics and Output

Post-fatigue analysis revealed reductions in GRF across all punches, with the greatest losses observed in the left hook (−68.86 N) and right uppercut (−61.17 N). Punch output also declined, particularly in the left hook (−8.21%), cross (−7.18%) and right hook (−6.25%), with an overall output reduction of 4.26% ($p = 0.027$). These results are consistent with prior findings that fatigue diminishes ground force production and disrupts kinetic chain transfer [19,20].

Interestingly, RFD did not uniformly decrease. The jab and right uppercut displayed increased RFD values post-fatigue (+255.67 N/s and +135.43 N/s, respectively), possibly reflecting a neuromuscular strategy to maintain punch velocity through faster muscle activation. This is in line with the force-velocity relationship, where diminished force capacity can be offset by increased contraction speed [18,21,22]. While potentially beneficial for preserving speed, these adjustments may compromise mechanical efficiency or elevate injury risk [23].

Differences in fatigue responses among punches indicate varied dependence on the lower limbs. These distinctions should guide fatigue-specific conditioning, particularly when structuring high-intensity rounds or sparring protocols [24,25]. Conditioning aimed at the cross and hooks may require enhanced focus on sustaining lower limb force output under fatigue.

4.3. Strong Versus Weak Punch Comparisons

This study's intra-individual comparison of the strongest versus the weakest punches offers unique insights. Across participants, weaker punches consistently underperformed. The left hook demonstrated a 22.8% performance gap with notable differences also seen in the left uppercut (−13.85%), right uppercut (−9.98%), jab (−9.31%) and right hook (−9.16%). These results emphasise that improving consistency across punch types may be as crucial as raising peak performance [25,26].

Although GRF and RFD data were only collected for the strongest punches, punch output data alone revealed meaningful discrepancies. These may stem from subtle differences in technique, neuromuscular readiness, or psychological focus. Steib et al. [23] noted that fatigue-related postural instability could affect the consistency of force application, reinforcing this interpretation.

These findings have practical coaching implications. Monitoring weak punch output-not just peak performance-could enhance training specificity, reduce variability and improve overall striking reliability [27].

4.4. Methodological Strengths and Limitations

This study employed high-frequency (1000 Hz) force plates and punch-specific force sensors to provide detailed kinetic analysis. Dividing punches into four phases, Base, Load, Drive and Impact, enabled precise segmentation of movement. The fatigue protocol was designed to replicate lower limb exertion typical of a boxing round, and the use of the Borg scale (RPE $\geq$ 18) ensured sufficient fatigue was reached.

The study did not incorporate electromyographic (EMG) measures, which limits the interpretation of underlying neuromuscular activation. However, the primary focus was on kinetic force-related variables, which were reliably captured using the available equipment. While the Hoteam device lacks formal lab calibration, it provided consistent relative output supporting its use for comparative force assessments in applied sport environments. This methodological trade-off reflects both the practical constraints and applied nature of the research setting. Other limitations were the small sample size (n = 10), which limits generalizability and statistical power. The small sample size (n = 10) inevitably reduces the statistical power of the study, increasing the likelihood of Type II error and limiting the ability to detect subtle or moderate effects. As a result, statistically non-significant findings should be interpreted cautiously, as they may reflect insufficient power rather than the absence of a true effect. Additionally, the limited number of participants restricts the generalizability of the results. With such a small cohort, individual variability may disproportionately influence group outcomes, and the findings may not fully represent the broader population of athletes or individuals with differing characteristics. While the study offers valuable preliminary insights, future research with larger, more diverse samples is necessary to confirm these patterns and strengthen external validity. The RFD results showed very high variability (i.e., large standard deviation). This may have been attributed to differences in technique among the boxers and/or the dynamic nature of the parameter.

Additionally, extensive post-processing was required: 960 punch phases were identified and segmented into 4800 subframes for averaging, requiring approximately 70 h of manual labour. While time-intensive, this enhanced data accuracy reinforced the robustness of the analysis. Future studies should expand sample size, include professional athletes and integrate motion capture and EMG data. This could improve understanding of trunk and lower limb coordination during punch execution under fatigue.

The Hoteam Boxing Force Sensor is a practical, wearable field device that enables ecologically valid testing of boxing punches. However, compared with laboratory-grade force plates and load cells, it has lower absolute accuracy and a narrower effective bandwidth. While we calibrated the sensor across the expected force range and demonstrated strong within-participant repeatability, direct comparisons with laboratory references yielded a small systematic bias and wider limits of agreement. Consequently, the Hoteam sensor is most appropriate for detecting relative changes (for example, pre- vs. post-fatigue) and between-condition comparisons rather than for obtaining gold-standard absolute force values. We minimized limitations through careful calibration and standardised sensor placement; however, consistency of gloves and hand wrapping across all participants was not controlled for. Participants were given the flexibility to use their own boxing gloves and hand wraps.

It must be acknowledged that using a vertical force plate integrated with the Vicon Motion System would have been preferred. However, a force plate, which is considered the gold standard, has inherent limitations [28,29] Its measurement error is not zero. Research has shown that errors increase particularly when the normal forces applied perpendicular to the plate's surface are small relative to the horizontal forces and/or when the point of force application is positioned near the edge of the plate.

4.5. Practical Applications and Future Research

The findings offer clear applications for boxing coaches and sports scientists. Crosses and hooks rely heavily on rear leg drive and should be targeted with strength and power training interventions, such as Olympic lifts, plyometrics, jump squats and resisted shadowboxing [6,18]. Fatigue resistance should also be developed through HIIT, circuit training and fatigued bag work [30–33].

Further practical implications include that the observed reductions in punch force and alterations in lower-limb kinetics following fatigue emphasize the need for training strategies that enhance the endurance capacity of the kinetic chain. Incorporating targeted lower-limb conditioning, technical drills performed under controlled fatigue, and systematic monitoring of force output may help preserve striking efficiency during high-intensity bouts. These findings underscore the importance of fatigue-informed training load management to optimize performance and minimize the risk of technique breakdown in competitive situations.

Identifying intra-individual performance gaps can guide technique refinement and reduce variability [26]. Future research should assess whether interventions like velocity-based or isometric training can attenuate fatigue-induced performance loss. Longitudinal studies (including predictive simulation studies) exploring adaptations to periodised programs aimed at improving rear leg output may provide further insights [34]. Motion capture and EMG should also be employed to deepen understanding of neuromuscular adaptations and kinetic chain sequencing-particularly for complex strikes like hooks and uppercuts [16].

5. Conclusions

This study examined the importance of lower limb kinetics in boxing punch performance and assessed the effects of fatigue on GRF, RFD and punch force across six punch types. Findings confirmed that rear leg drive is critical to producing high punch output, particularly in rotational strikes such as the cross and hook. Fatigue significantly reduced punch force and GRF, while RFD showed variable responses, with some punches demonstrating compensatory increases. Intra-individual comparisons revealed that weaker punches experienced greater performance declines, emphasising the importance of training for consistency as well as peak output. These insights highlight the value of targeted lower limb conditioning and fatigue-specific strategies to preserve kinetic chain efficiency. Future research should incorporate larger samples, EMG and motion capture to enhance understanding of neuromuscular coordination during fatigued striking.

Author Contributions: Conceptualization, U.C.U. and C.S.; methodology, U.C.U., C.S. and R.C.; software, U.C.U.; validation, U.C.U., C.S. and R.C.; formal analysis, U.C.U. and C.S.; investigation, U.C.U., C.S. and J.S.B.; resources, U.C.U.; data curation, U.C.U., C.S., J.S.B. and Y.G.; writing—original draft preparation, C.S. and U.C.U.; writing—review and editing, U.C.U., C.S., J.S.B., Y.G. and F.D.; visualization, U.C.U. and C.S.; supervision, U.C.U.; project administration, U.C.U. and Y.G. All authors have read and agreed to the published version of the manuscript.

Funding: This research received no external funding.

Institutional Review Board Statement: The study was conducted in accordance with the Declaration of Helsinki, and approved by the Ethics Committee of the University of the West of Scotland, UK (Approval Number: 15871 and Date of Approval: 4 November 2022).

Informed Consent Statement: Informed consent was obtained from all subjects involved in the study.

Data Availability Statement: The data that support the findings of this study are not openly available due to reasons of sensitivity and are available from the corresponding author upon reasonable request. Data are located in controlled access data storage at the University of the West of Scotland.

Acknowledgments: The authors would like to thank Rachel Stewart and Favour Igoli for running earlier pilot studies to test the efficacy of our new laboratory-based boxing protocols.

Conflicts of Interest: The authors declare no conflicts of interest.

References

1. Da Broi, M.; Al Awadhi, A.; Voruz, P.; Nouri, A.; Schaller, K. The spectrum of acute and chronic consequences of neurotrauma in professional and amateur boxing: A call to action is advocated to better understand and prevent this phenomenon. *Brain Spine* **2023**, *4*, 102743. [CrossRef] [PubMed]
2. Donnelly, R.R.; Ugbolue, U.C.; Gao, Y.; Gu, Y.; Dutheil, F.; Baker, J.S. A systematic review and meta-analysis investigating head trauma in boxing. *Clin. J. Sport Med.* **2023**, *33*, 658–674. [CrossRef] [PubMed]
3. McGill, S.M.; Chaimberg, J.D.; Frost, D.M.; Fenwick, C.M.J. Evidence of a double peak in muscle activation to enhance strike speed and force: An example with elite mixed martial arts fighters. *J. Strength Cond. Res.* **2010**, *24*, 348–357. [CrossRef] [PubMed]
4. Whiting, W.C.; Gregor, R.J.; Finerman, G.A. Kinematic analysis of human upper extremity movements in boxing. *Am. J. Sports Med.* **1988**, *16*, 130–136. [CrossRef]
5. Zazryn, T.; Cameron, P.; McCroy, P. A prospective cohort study of injury in amateur and professional boxing. *Br. J. Sports Med.* **2006**, *40*, 670–674. [CrossRef]
6. Dunn, E.C.; Humberstone, C.E.; Franchini, E.; Iredale, K.F. The effect of fatiguing lower-body exercise on punch forces in highly trained boxers. *Eur. J. Sport Sci.* **2022**, *22*, 964–972. [CrossRef]
7. Stanley, E.; Thomson, E.; Smith, G.; Lamb, K.L. An analysis of the three-dimensional kinetics and kinematics of maximal effort punches among amateur boxers. *Int. J. Perform. Anal. Sport* **2018**, *18*, 835–854. [CrossRef]
8. Lenetsky, S.; Brughelli, M.; Nates, R.J.; Neville, J.G.; Cross, M.R.; Lormier, A.V. Defining the phases of boxing punches: A mixed-method approach. *J. Strength Cond. Res.* **2020**, *34*, 1040–1051. [CrossRef]
9. Dinu, D.; Louis, J. Biomechanical analysis of the cross, hook, and uppercut in junior vs. elite boxers: Implications for training and talent identification. *Front. Sports Act. Living* **2020**, *2*, 598861. [CrossRef]
10. Liu, Y.; Wang, Y.; Zhang, J.; Zhang, Y. Biomechanics of the lead straight punch of different level boxers. *Front. Physiol.* **2022**, *13*, 1015154. [CrossRef]
11. Perlinski, J.; Sikorski, W.; Wochna, K.; Raczek, N. Gait analysis of male professional boxers. *Balt. J. Health Phys. Act.* **2024**, *16*, 1. [CrossRef]
12. Menzel, T.; Potthast, W. Validation of a novel boxing monitoring system to detect and analyse the centre of pressure movement on the boxer's fist. *Sensors* **2021**, *21*, 8394. [CrossRef] [PubMed]
13. Boxing Strength Tester-Punch Force Sensor | Equipment Wall Mounted Boxing Boxer | Tester Boxing Training Sandbag Vent Target | Wall Punch Pad Boxing Punching Bag Machine for Adult Kid: Amazon.co.uk: Sports & Outdoors. Available online: https://www.amazon.co.uk/Boxing-Strength-Tester-Equipment-Training/dp/B0FJPS79Y2/ref=sr_1_2?crid=33AQUHWQJXOEJ&dib=eyJ2IjoiMSJ9.xr1DY1Qpz9O4fZc0l2laFUVZ6VR8SDLaa3SlrWnS7fRrtQT80iA8V-bWRrnTYVkpoFiTCXqm8q3dMXB6RoTxAwT0wrhWVPZBOUWtz1Gi4ZwqVPbdaFH-AwtfzcqN5uiDtC4x8ryk7WNYQgL-AdlnmHwDjBWqsS5LVj7Nn6gjP520xFXygpaweh56r48Bbd9UIg-8A3gvNYe8O_lV3wYkqd0q6kM5t4Kg7mj8n79Abwb4hkV3T7ekLxuZztdbaTmH8VAAOgyHAib72sQ3ZzakZW4aqS0kWuwbkqklljb0Ep8.QFzOS3vF-eQhDuW2wHxNoLB4Kk0hMipqWJO-Qj_6LHE&dib_tag=se&keywords=boxing+punch+tester&qid=1764391337&s=sports&sprefix=boxing+punch+tester,sports,210&sr=1-2 (accessed on 26 November 2025).
14. Sayyadi, P.; Razeghi, M.; Jalali, M. The effectiveness of fatigue on repositioning sense of lower extremities: Systematic review and meta-analysis. *BMC Sports Sci. Med. Rehabil.* **2024**, *16*, 35. [CrossRef] [PubMed]
15. Cohen, J. *Statistical Power Analysis for the Behavioral Sciences*; Routledge: Abingdon, UK, 2013.
16. Almansoof, B.; Batal, M.; Youm, Y. Development of a musculoskeletal model for lower limb motion analysis during boxing punches. *Procedia Comput. Sci.* **2021**, *179*, 218–224. [CrossRef]
17. Yi, W.; Chen, C.; Zhou, Z.; Cui, W.; Wang, D. Acute effects of ballistic versus heavy-resistance exercises on countermovement jump and rear-hand straight punch performance in amateur boxers. *BMC Sports Sci. Med. Rehabil.* **2022**, *14*, 161. [CrossRef]
18. Loturco, I.; Artioli, G.G.; Kobal, R.; Gil, S.; Franchini, E.; Nakamura, F.Y. Strength and power qualities are highly associated with punching impact in elite amateur boxers. *J. Strength Cond. Res.* **2016**, *30*, 109–116. [CrossRef]

19. Tong-Iam, R.; Rachanavy, P.; Lawsirirat, C. Kinematic and kinetic analysis of throwing a straight punch: The role of trunk rotation in delivering a powerful straight punch. *J. Phys. Educ. Sport* **2017**, *17*, 2538–2543.
20. Haralabidis, N.; Saxby, D.J.; Pizzolato, C.; Needham, L.; Cazzola, D.; Minahan, C. Fusing Accelerometry with Videography to Monitor the Effect of Fatigue on Punching Performance in Elite Boxers. *Sensors* **2020**, *20*, 5749. [CrossRef]
21. Cormie, P.; McGuigan, M.R.; Newton, R.U. Developing maximal neuromuscular power: Part 1—Biological basis of maximal power production. *Sports Med.* **2011**, *41*, 17–38. [CrossRef]
22. Cormie, P.; McGuigan, M.R.; Newton, R.U. Developing maximal neuromuscular power: Part 2—Training considerations for improving maximal power production. *Sports Med.* **2011**, *41*, 125–146. [CrossRef]
23. Steib, S.; Hentschke, C.; Welsch, G.; Pfeifer, K.; Zech, A. Effects of fatiguing treadmill running on sensorimotor control in athletes with and without functional ankle instability. *Clin. Biomech.* **2013**, *28*, 790–795. [CrossRef]
24. Sánchez-Ramírez, C.; Cid-Calfucura, I.; Hernandez-Martinez, J.; Cancino-López, J.; Aedo-Muñoz, E.; Valdés-Badilla, P.; Franchini, E.; García-García, J.M.; Calvo-Rico, B.; Abián-Vicén, J.; et al. Submaximal Accentuated Eccentric Jump Training Improves Punching Performance and Countermovement Jump Force–Time Variables in Amateur Boxers. *Appl. Sci.* **2025**, *15*, 7873. [CrossRef]
25. Dunn, E.C.; Humberstone, C.E.; Franchini, E.; Iredale, K.F.; Blazevich, A.J. Relationships Between Punch Impact Force and Upper- and Lower-Body Muscular Strength and Power in Highly Trained Amateur Boxers. *J. Strength Cond. Res.* **2022**, *36*, 1019–1025. [CrossRef] [PubMed]
26. Mosler, D.; Kacprzak, J.; Wąsik, J. The influence of effective mass on the striking force of lead jab and rear cross punches of boxers. *Appl. Sci.* **2024**, *14*, 7785. [CrossRef]
27. Lenetsky, S.; Brughelli, M.; Nates, R.J.; Cross, M.R.; Lormier, A.V. Variability and Reliability of Punching Impact Kinetics in Untrained Participants and Experienced Boxers. *J. Strength Cond. Res.* **2018**, *32*, 1838–1842. [CrossRef]
28. Schmiedmayer, H.B.; Kastner, J. Parameters influencing the accuracy of the point of force application determined with piezoelectric force plates. *J. Biomech.* **1999**, *32*, 1237–1242. [CrossRef]
29. Chesnin, K.J.; Selby-Silverstein, L.; Besser, M.P. Comparison of an in-shoe pressure measurement device to a force plate: Concurrrent validity of center of pressure measurements. *Gait Posture* **2000**, *12*, 128–133. [CrossRef]
30. Herrera-Valenzuela, T.; Carter, J.; Leiva, E.; Valdés-Badilla, P.; Ojeda-Aravena, A.; Franchini, E. Effect of a short HIIT program with specific techniques on physical condition and activity during simulated combat in national-level boxers. *Sustainability* **2021**, *13*, 8746. [CrossRef]
31. Usher, A.; Babraj, J. Impact of sprint interval training on post-fatigue mitochondrial rate in professional boxers. *Eur. J. Appl. Physiol.* **2025**, *125*, 261–271. [CrossRef]
32. Nukeaw, A.; Boonrod, W. Effect of Supplemented Muay Thai Circuit Training Program on Maximal Oxigen Uptake and Physical Performance in Professional Muay Thai Boxers. *J. Sports Sci. Health* **2020**, *21*, 355–367.
33. Niu, Z.; Huang, Z.; Zhao, G.; Chen, C. Impact of three weeks of integrative neuromuscular training on the athletic performance of elite female boxers. *PeerJ* **2024**, *12*, e18311. [CrossRef]
34. Haralabidis, N.; Colyer, S.L.; Serrancolí, G.; Salo, A.I.T.; Cazzola, D. Modifications to the net knee moments lead to the greatest improvements in accelerative sprinting performance: A predictive simulation study. *Sci. Rep.* **2022**, *12*, 15908. [CrossRef]

Disclaimer/Publisher's Note: The statements, opinions and data contained in all publications are solely those of the individual author(s) and contributor(s) and not of MDPI and/or the editor(s). MDPI and/or the editor(s) disclaim responsibility for any injury to people or property resulting from any ideas, methods, instructions or products referred to in the content.

bioengineering

MDPI

Article

Assessment of Muscle Synergies in Chronic Ankle Instability Patients During Unanticipated and Anticipated Landing

Zhifeng Zhou [1], Datao Xu [1], Meizi Wang [1,2], Tianle Jie [1], Julien S. Baker [1], Huiyu Zhou [1] and Yaodong Gu [1,3,*]

1 Faculty of Sports Science, Ningbo University, Ningbo 315211, China; zhouzhifeng_nbu@outlook.com (Z.Z.); xudatao3@gmail.com (D.X.); meiziwang@polyu.edu.hk (M.W.); jietianle_nbu@outlook.com (T.J.); jsbaker@hkbu.edu.hk (J.S.B.); zhouhuiyu@nbu.edu.cn (H.Z.)

2 Department of Biomedical Engineering, Faculty of Engineering, The Hong Kong Polytechnic University, Kowloon, Hong Kong SAR 999077, China

3 Faculty of Engineering, University of Szeged, 6720 Szeged, Hungary

* Correspondence: guyaodong@nbu.edu.cn; Tel.: +86-574-87600208

Abstract: Ankle sprains are a common injury among athletes and the general population, with chronic ankle instability (CAI) being a frequent complication. CAI patients often display altered neuromuscular control adaptations. This study analyzed muscle synergy patterns in 20 CAI patients during anticipated and unanticipated landing tasks to understand their neuromuscular adaptation strategies. Using Nesterov non-negative matrix factorization and K-means clustering, the study identified distinct muscle activation patterns. Results indicated that during unanticipated landings, the gluteus maximus and vastus lateralis showed increased activation weight, while the medial gastrocnemius was more active in anticipated landings. This study highlights that CAI patients display unique muscle synergy patterns during unanticipated landings, relying more on proximal muscles such as the gluteus maximus and vastus lateralis. This adaptation reflects the proximal muscle strategy to enhance stability and compensate for impaired ankle function in unpredictable situations.

Keywords: neuromuscular activation; muscle synergy; non-negative matrix factorization; unanticipated landing; chronic ankle instability

1. Introduction

Ankle injuries, particularly ankle sprains, are common in athletes and the general population during everyday activities. These injuries not only cause acute pain and functional impairment but can also lead to chronic issues, such as chronic ankle instability (CAI) [1]. CAI is characterized by recurring ankle sprains and persistent feelings of instability following an initial ankle sprain. Epidemiological studies indicate that approximately 40% to 70% of individuals with an ankle sprain may develop CAI. CAI not only impacts athletic performance but also increases the risk of future injuries and can result in long-term functional impairments [2]. The complexity of ankle injuries lies in their involvement of multiple factors beyond a single structural issue. This includes proprioceptive loss, altered neuromuscular control, and coordination of motor skills. Ankle instability is often associated with a loss of proprioception and changes in neuromuscular control strategies, with these factors collectively affecting joint stability and function [3].

In studying the movement strategies of chronic ankle instability (CAI) patients, unanticipated cutting tasks are a critical focus. These tasks often lead to marked alterations in movement strategies, particularly when comparing anticipated versus unanticipated conditions. Unanticipated tasks place greater demands on performance, requiring increased muscle activation and more complex movement control strategies [4]. Such tasks typically require athletes to respond rapidly in uncertain environments, challenging their dynamic stability and motor coordination [5]. Studies indicate that CAI patients often display different movement strategies compared to healthy individuals during unanticipated cutting tasks. For example, Kim's research found that CAI patients rely more heavily on tibialis anterior activation during both the early and late stages of landing and cutting tasks [6]. This adjustment serves as a compensatory strategy for the inadequacies of ankle joint function. Moreover, unanticipated landing tasks not only reveal deficiencies in CAI patients' movement control but also highlight their adaptability in managing sudden situations [7]. These adaptations may involve cognitive processing of movement tasks, adjustments in neuromuscular control strategies, and responses to unstable environments [8]. Compared to healthy individuals, CAI patients may show greater movement instability and slower adaptability during unanticipated tasks, potentially increasing their risk of injury in real-world activities. Specifically, these patients tend to rely more on proximal muscles, such as the gluteus maximus and vastus lateralis, to maintain overall stability during landings [9]. Ankle injuries lead to adaptive changes in the central nervous system, which affect motor control. Consequently, proximal joints of the injured lower limb may exhibit neuromuscular control deficits. Thus, investigating movement patterns during unanticipated landing tasks is vital to creating efficient rehabilitation strategies and interventions [10].

Muscle synergy patterns are a central concept in the study of neuromuscular control and play a crucial role in motor control research. Initially proposed by Bernstein, muscle synergies refer to groups of "cooperating" muscle groups. This concept has evolved significantly, especially with advancements in non-negative matrix factorization (NNMF) algorithms [11]. NNMF simplifies complex multidimensional data by decomposing muscle activation signals into several time-invariant synergy modules, allowing for detailed analysis through specific task activation coefficients [12]. Building upon this process, the NNMF algorithm identifies modules associated with time-invariant muscle synergy patterns from the input muscle activation matrix and generates activation coefficients for individual tasks [13]. This decomposition method enables researchers to identify and quantify the synergistic actions of muscle groups during movement, revealing their functions in different motor tasks. The advantage of this technique lies in its non-negativity constraint on the original data, which preserves the true characteristics of the data while accurately capturing the core patterns of muscle activity [14,15]. In CAI research, NNMF algorithms have been widely applied. Studies have shown that NNMF can provide a detailed analysis of muscle synergy patterns in CAI patients across various motor tasks, helping to identify key muscle groups and their synergies associated with CAI [16,17]. For instance, Ghislieri et al. found that CAI patients exhibited poorer balance in single-leg tasks compared to healthy controls and demonstrated reduced muscle synergy in the absence of visual information. This indicates a high reliance on visual input to compensate for proprioceptive deficits affecting the ankle joint [18].

Therefore, this study aims to assess and compare muscle synergy patterns in patients with CAI during anticipated and unanticipated landing tasks. We hypothesize that CAI patients will exhibit significant changes in their muscle-specific functional roles during unanticipated landings, which will better reflect their control strategies when faced with unexpected situations compared to anticipated landings. By analyzing the muscle activity of CAI patients in these two different contexts, we hope to reveal the neuromuscular coordination capabilities and adaptive responses required for effective movement execution. Furthermore, the findings of this research will provide important theoretical support for understanding CAI patients' ability to maintain stability and balance in dynamic environments, thereby guiding clinical rehabilitation interventions to improve patients' athletic performance and quality of life.

2. Materials and Methods

2.1. Participant

In this study, 20 participants with chronic ankle instability (CAI) were recruited. Each participant completed the self-administered Cumberland Ankle Instability Tool (CAIT) questionnaire, which gathered information on anthropometric parameters, the frequency of lateral ankle sprains, and their history of medical consultations. According to the criteria set by the International Ankle Consortium, CAI is defined by the following: (1) a CAIT score of 24 or lower, (2) a history of traumatic ankle sprains requiring medical attention on two or more occasions, and (3) recurrent lateral ankle sprains lasting for at least 6 months or occurring frequently. To account for the potential impact of limb dominance on the experiment, all participants' right legs were designated as the dominant limb, defined as the preferred leg for kicking a ball. Participants were fully briefed on the study's objectives, requirements, and procedures, and provided their written informed consent. The study protocol was approved by the Research Ethics Committee of Ningbo University (Approval Number: RAGH20240622).

2.2. Experimental Protocol and Data Recordings

This study employed a Kistler force platform alongside an eight-camera Vicon motion capture system from Oxford Metrics Ltd. (Oxford, UK) to simultaneously capture dynamic and kinematic data. Kinematic data were acquired at a frequency of 200 Hz, while dynamic data were collected at 1000 Hz. Mark points were set according to the 2392 generic model, and EMG sensors were positioned following SENIAM guidelines [19]. To prepare the measurement area, hair was shaved from the skin and the area was wiped with alcohol to minimize the impedance at the skin-electrode interface. Eight EMG sensors (Delsys, Boston, MA, USA) were placed on the muscle belly of the soleus (SL), medial gastrocnemius (MG), lateral gastrocnemius (LG), tibialis anterior (TA), peroneus longus (PL), rectus femoris (RF), vastus medialis (VM), and vastus lateralis (VL) to quantify muscle activation (Figure 1A).

Before the formal experiment, each movement was preceded by adequate warm-up exercises. During the warm-up phase, participants first familiarized themselves with the task procedure and landed several times from a two-step platform to ensure they had mastered the correct landing techniques. In the subsequent unanticipated landing task, participants were required to descend from two steps, each 20 cm high, and land on a landing board with their right leg. They were required to remain on the force platform until their foot touched the ground. Participants were unaware of whether the springboard was tilted. A sponge was placed under the springboard; and when the foot made contact, the sponge collapsed, causing the board to tilt. All tasks were performed barefoot, and each participant was required to complete 5 trials of the unanticipated and anticipated landing task successfully. The order of the anticipated and unanticipated landing tasks was alternated in a randomized sequence to simulate unexpected conditions and ensure the non-prepared nature of the unanticipated landings. For analysis purposes, the results were divided into 'anticipated' and 'unanticipated' groups based on the specific task condition performed by the participants during the trial. The unanticipated trial was considered successful if the entire foot contacted the ground, and the trial was deemed unsuccessful if the participant could not maintain balance. Only trials that were considered successful were included in the analysis.

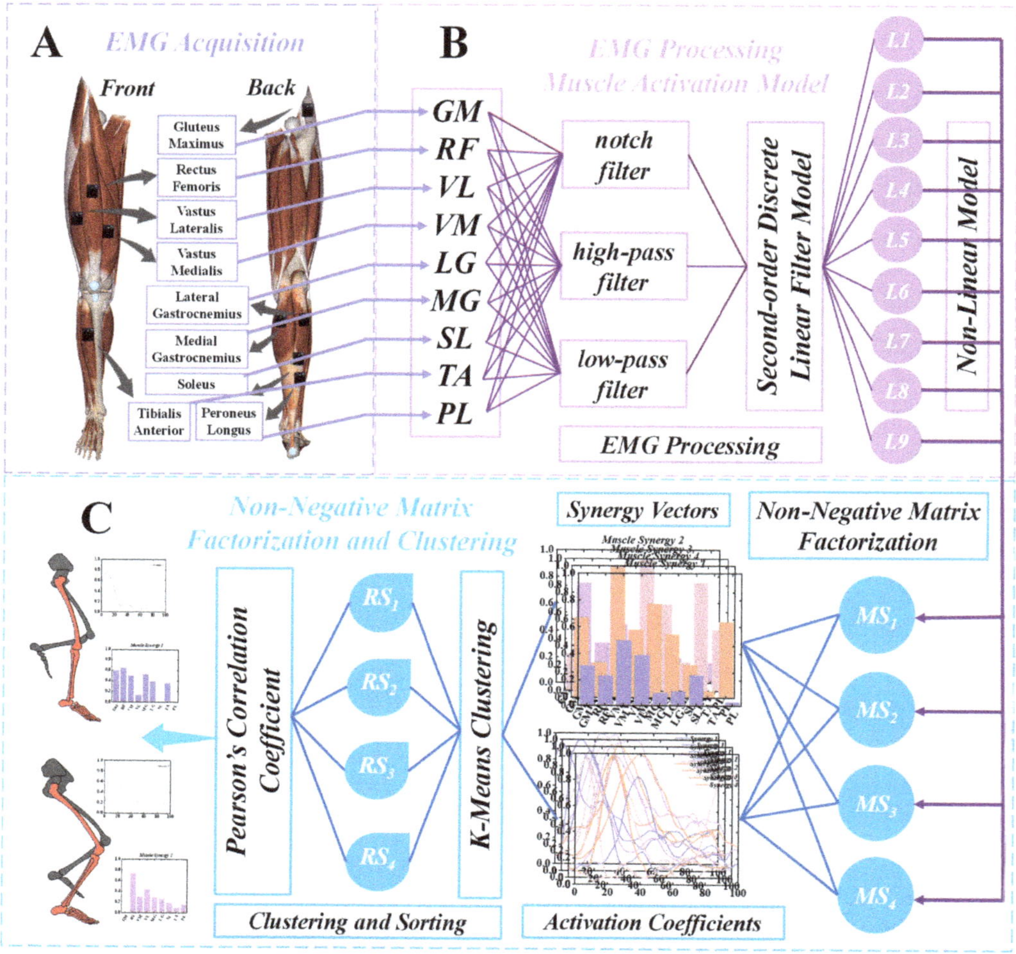

Figure 1. The overall workflow of the current study (**A**) Acquisition of sEMG. (**B**) Detailed Processing for sEMG data. (**C**) Muscle synergy extraction using the NNMF algorithm and Sorting of muscle synergies using the K-means clustering algorithm.

2.3. Data Processing

This study utilized Vicon Nexus 2.14.0 software (Vicon Metrics Ltd., version 2.14.0, Oxford, UK) to identify and acquire three-dimensional marker trajectories and ground reaction force (GRF) data, which were exported in C3D format. Concurrently, a custom MATLAB 2022b (MathWorks, Inc., Natick, MA, USA) was employed to construct a muscle activation model from the raw surface electromyography (EMG) signals. Data analysis focused on the posture phases of each task. For the landing task, a GRF threshold of 10 N was established to define the posture phase. This threshold was set to accurately capture the stability and mechanical characteristics during the landing process. The EMG data underwent a detailed processing sequence. Initially, a low-pass filter with a cutoff frequency of 450 Hz was applied to remove high-frequency noise and interference, while a high-pass filter with a cutoff frequency of 20 Hz was used to eliminate low-frequency drift and artifacts. This preliminary filtering step aimed to clean the signal and retain meaningful electromyographic information. Following filtering, the EMG data were rectified, converting all negative values to positive. This step facilitates subsequent analysis by ensuring that the signal is uniformly positive. The rectified data were then smoothed using a low-pass filter with a cutoff frequency of 10 Hz, which reduced high-frequency fluctuations and produced a more stable signal suitable for analysis. Finally, the smoothed EMG signals were normalized using peak normalization [20]. This process standardized each signal by scaling the maximum value to a common reference, thus removing individual differences between trials or participants and enabling accurate comparisons under different conditions (Figure 1B).

Through these meticulous processing steps, we obtained normalized EMG signals that provided a reliable foundation for further analysis of muscle activation patterns. This approach ensured a thorough examination of the posture phases, offering insights into the relationship between muscle activation and movement performance (Figure 2).

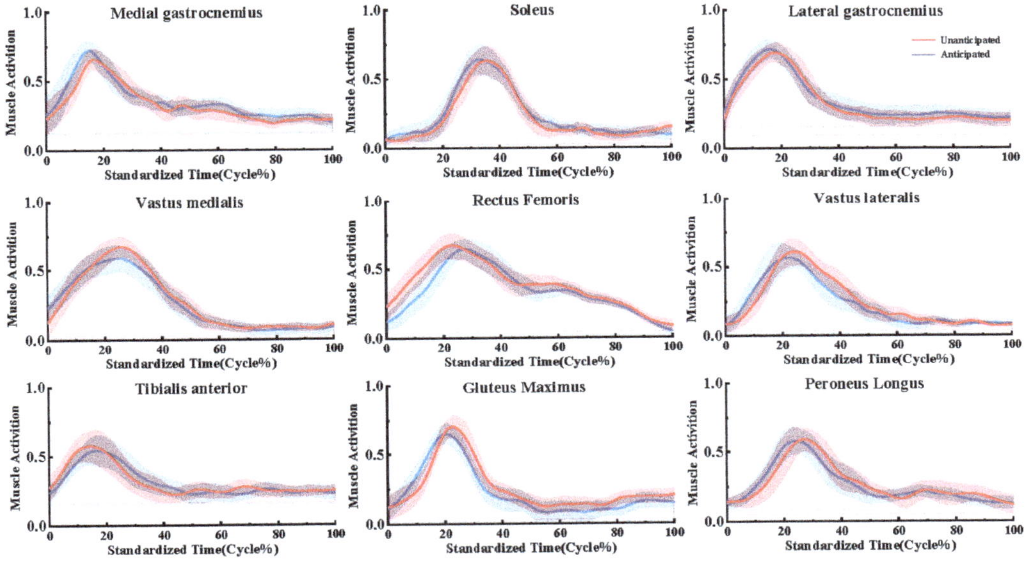

Figure 2. Muscle activation profiles during landing in the Unanticipated and Anticipated groups.

2.4. Nesterov Non-Negative Matrix Factorization Extracts Muscle Synergies

Non-negative matrix factorization (NNMF) is a method that exploits the intrinsic relationships within data to derive a mapping matrix. By projecting high-dimensional data onto a lower-dimensional subspace, NNMF is an effective tool for data clustering and dimensionality reduction tasks (Figure 1C). Specifically, NNMF decomposes high-dimensional data into the product of two lower-dimensional matrices. However, NNMF encounters certain efficiency issues during the decomposition process. Therefore, this paper adopts a novel approach, NeNMF, to decompose electromyography (EMG) signals. By utilizing NeNMF, we aim to enhance the efficiency and effectiveness of the decomposition, allowing for better representation of the underlying patterns in the EMG data and facilitating improved analysis and interpretation. The EMG data matrix V_{mn} is of dimensions $m \times n$, where m denotes the number of muscles and n denotes the number of data points [21]. The underlying principle of NeNMF is described as follows:

$$V_{mn} \approx (WH)_{mn} = \sum_{i=1}^{k} W_{mi}H_{in} = V'_{mn} \quad (1)$$

To approximate the decomposition $V \approx WH$, we first need to define a cost function to quantify this approximation [22]. According to Equation (1), and starting from an initial

$W^1 \geq 0$, the block coordinate descent method iteratively alternates between solving the subproblems defined in the optimization framework, the function can be defined as follows:

$$H^{t+1} = \arg \min_{H \geq 0} F(W^t, H) = \frac{1}{2} \| X - W^t H \|_F^2 \tag{2}$$

After optimization, we obtain the following:

$$W^{t+1} = \arg \min_{W \geq 0} F(W, H^{t+1}) = \frac{1}{2} \| X^T - H^{t+1} W^T \|_F^2 \tag{3}$$

This study uses Nesterov's Optimal Gradient Method (OGM) to optimize the NNMF process. By applying Nesterov's approach, we aim to increase both the speed and accuracy of convergence, leading to more efficient data analysis, such as better feature extraction and representation.

In the realm of optimization, recent findings indicate that smooth optimization methods based on gradients can lead to improved convergence rates. Because the gradient satisfies Lipschitz continuity, we can effectively apply Nesterov's method to optimize Equations (2) and (3). To facilitate this process, we define two sequences that are updated during each iteration, and these sequences are described as follows:

$$\begin{aligned} H_k &= \arg \min_{H \geq 0} \phi(Y_k, H) = F(W^t, Y_k) \\ &+ \langle \nabla_H F(W^t, Y_k), II - Y_k \rangle + \tfrac{L}{2} \| II - Y_k \|_F^2 \end{aligned} \tag{4}$$

And

$$Y_{k+1} = II_k + \frac{\alpha_k - 1}{\alpha_{k+1}} (II_k - II_{k-1}) \tag{5}$$

Previous research indicates that the optimized NeNMF method offers significant advantages over NNMF solvers. Specifically, NeNMF maintains a comparable computational complexity while achieving faster convergence rates in each iteration. This efficiency allows for quicker results, making it particularly beneficial for applications that require timely matrix decomposition. Additionally, the optimized NeNMF effectively captures the underlying structures in the data, ensuring accurate representation and interpretation.

Additionally, the Variance Accounted For (VAF) value indicates the extent to which the global muscle activity (VAF_{global}) and the local muscle activity (VAF_{muscle}) in the electromyography (EMG) data can be explained by their respective number of synergies:

$$\text{VAF}_{global} = 1 - \frac{\sum_{m,n} (V - V_r)^2_{m,n}}{\sum_{m,n} V^2_{m,n}} \tag{6}$$

$$\text{VAF}_{muscle} = 1 - \frac{\sum_n (V - V_r)^2_{m_{each} n}}{\sum_n V^2_{m_{each} n}} \tag{7}$$

At the same time, the number of synergies must meet the following criteria: the global muscle variance accounted for (VAF_{global}) should be $\geq$90%, and the local muscle variance accounted for (VAF_{muscle}) should be $\geq$75% [23].

2.5. K-Means Clustering Algorithm for Sorting Muscle Synergies

In this study, we employed the K-means clustering algorithm to classify muscle synergies. First, we organized the muscle activity data into a matrix X. After standardizing the data, we applied the K-means algorithm to identify different muscle synergy patterns [24]. The initial cluster centers μ_i were chosen randomly and optimized through the following steps:

$$\text{assign}(x_j) = \arg \min_i \| x_j - \mu_i \|^2 \tag{8}$$

Each cluster center μ_i to be the mean of all data points in the cluster:

$$\mu_i = \frac{1}{|C_i|} \sum_{x \in C_i} x \tag{9}$$

The assignment and update steps continue until the change in cluster centers is less than a specified threshold ϵ or until the maximum number of iterations, max_iters is reached. The change in cluster centers is calculated as follows:

$$\Delta\mu_i = \| \mu_i^{\text{new}} - \mu_i^{\text{old}} \| \tag{10}$$

After completing the K-means clustering, we further refined the classification of each subject's synergy vectors by calculating the Pearson correlation coefficient r_{ij} between each synergy vector W_j and each cluster center μ_i [25]. The Pearson correlation coefficient is computed as:

$$r_{ij} = \frac{\text{Cov}(W_j, \mu_i)}{\sigma_{W_j}\sigma_{\mu_i}} \tag{11}$$

where $\text{Cov}(W_j, \mu_i)$ is the covariance between the synergy vector and the cluster center, and σ_{W_j} and σ_{μ_i} are their respective standard deviations. If r_{ij} is greater than 0.6, we consider the synergy vector W_j to be highly similar to the cluster center μ_i and assign W_j to cluster i. Once the synergy vectors have been classified, their corresponding activation coefficients C are automatically grouped into the same category [26]. This method, combining K-means clustering with Pearson correlation, provides a refined classification framework that helps in analyzing different muscle synergy patterns and their variations under different conditions.

2.6. Statistical Analysis

The study commenced using statistical analysis employing the Shapiro–Wilk test to evaluate the normality of the data. If the data met the criteria for normal distribution, we proceeded with a paired *t*-test to analyze the differences between the two groups [27]. This approach allows us to discern whether there are substantial differences between the means of the two related samples, thereby shedding light on any significant effects or differences between the groups [28]. The formula for the paired *t*-test is given by:

$$t = \frac{\bar{d} - \mu_0}{\frac{s_d}{\sqrt{n}}} \tag{12}$$

In which,

$$\bar{d} = \frac{\sum_{i=1}^{n} d_i}{n} \tag{13}$$

In the paired *t*-test, this refers to the mean of the differences between paired samples:

$$s_d = \sqrt{\frac{\sum_{i=1}^{n} \left(d_i - \bar{d}\right)^2}{n-1}} \tag{14}$$

It is the standard deviation of the differences between paired samples, and n represents the number of paired samples.

To study the movement patterns of CAI patients' lower limbs with Statistical Parametric Mapping (SPM), we began by extracting and preparing the dataset. We used a MATLAB script to convert stance phase data into a time series of 101 points. Following this, we conducted the SPM analysis by applying thresholds derived from random field theory and creating the necessary SPM curves [29]. If SPM values go beyond the critical thresholds, it shows that there are significant differences in activation coefficients between the groups at that specific time, revealing important variations in muscle activity [30]. This method helps identify statistically significant differences within the time series, allowing for a more accurate analysis of the movement patterns in CAI patients.

3. Results

3.1. Muscle Synergy Number via NeNNMF Decomposition

This study investigates the muscle synergy patterns in patients with chronic ankle instability (CAI) during unanticipated and anticipated landing tasks. The analysis reveals that most CAI patients exhibit four primary muscle synergy patterns under both landing conditions. These patterns highlight the coordination and variability of muscle interactions in CAI patients across different landing scenarios. To improve clarity in our reporting, we focus exclusively on presenting the coordination vector matrices and activation coefficient matrices for the four primary muscle synergy patterns extracted from each subject group (Figures 3–6). These matrices visually represent the muscle synergies under various landing conditions. Specifically, the coordination vector matrices illustrate the relationships between different muscles, while the activation coefficient matrices reflect the intensity and patterns of muscle activation during the landing process.

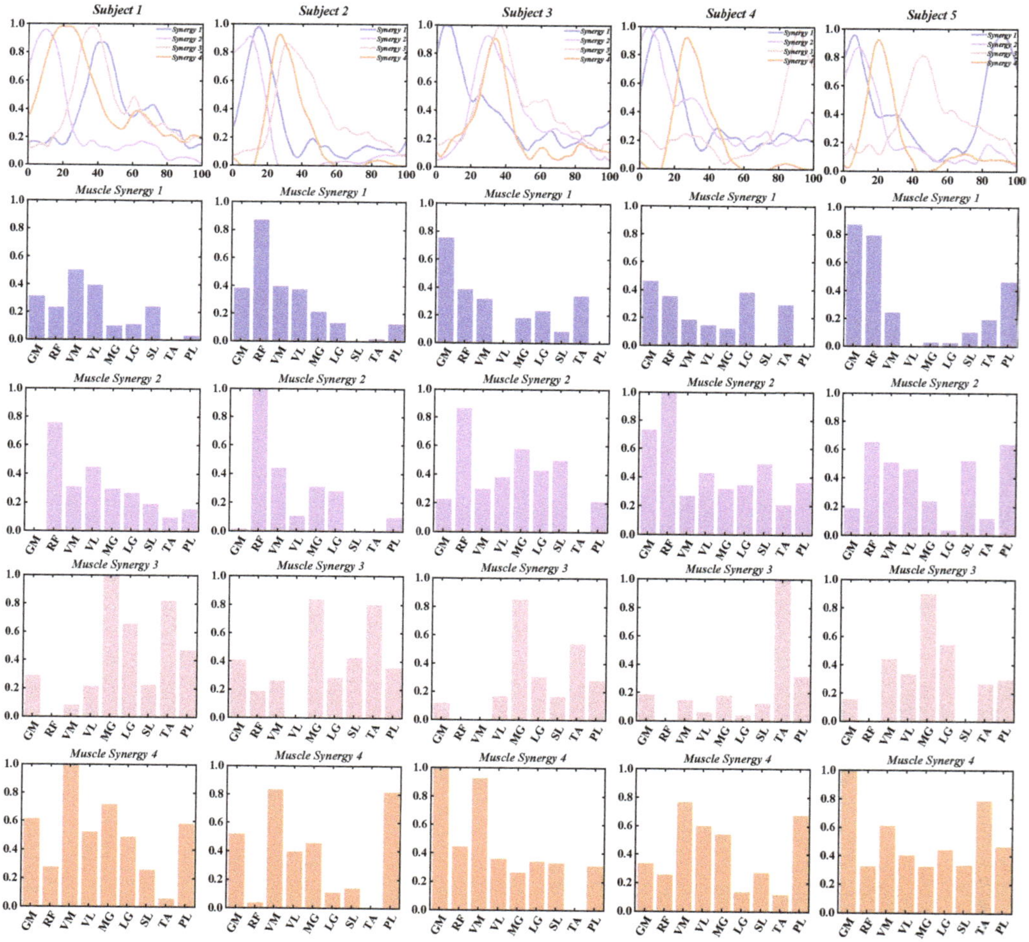

Figure 3. Detailed Presentation of activation coefficient results and synergy vectors for Subjects 1–5 in the Unanticipated Landing Group.

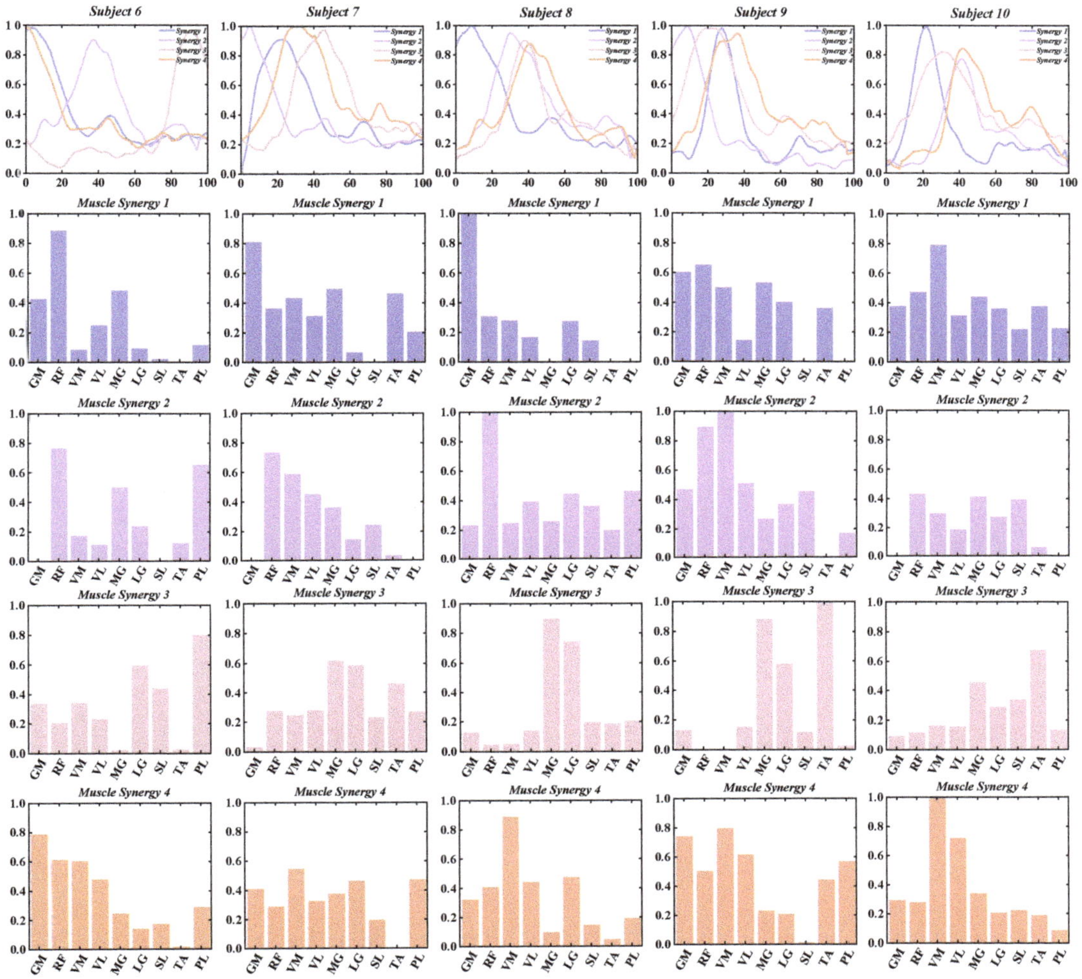

Figure 4. Detailed Presentation of activation coefficient results and synergy vectors for Subjects 6–10 in the Unanticipated Landing Group.

Based on the results from the Variance Accounted For (VAF) analysis, we found that patients with chronic ankle instability (CAI) demonstrated high levels of global VAF all muscle synergy patterns during both unanticipated and anticipated landing tasks (Figure 7).

In the unanticipated landing group, the global VAF values were as follows: 84.9% ± 1.5% with three muscle synergies, 90.0% ± 0.9% with four muscle synergies, and 92.4% ± 0.8% with five muscle synergies. Similarly, in the anticipated landing group, the global VAF values were: 85.3% ± 1.5% with three muscle synergies, 90.5% ± 1.2% with four muscle synergies, and 92.7% ± 0.9% with five muscle synergies (Table 1).

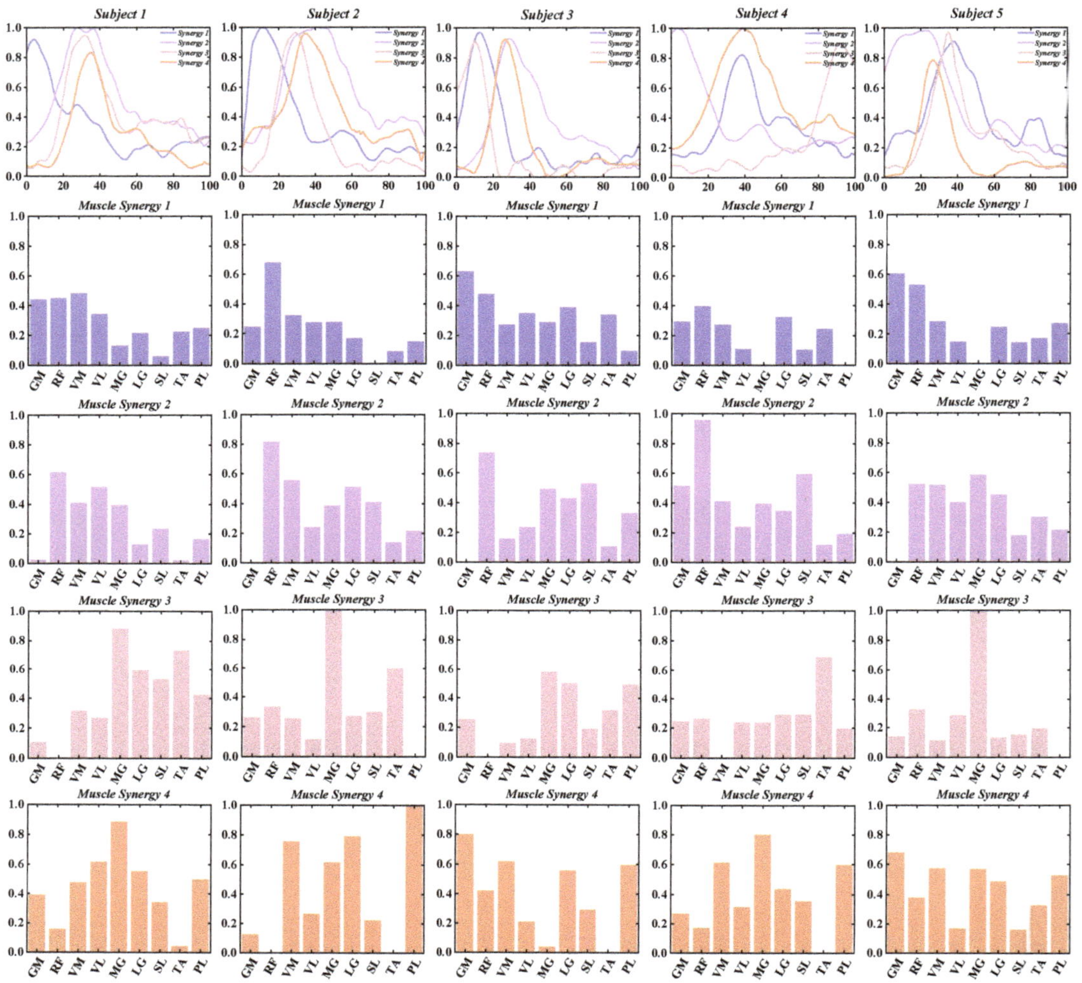

Figure 5. Detailed Presentation of activation coefficient results and synergy vectors for Subjects 1–5 in the Anticipated Landing Group.

Table 1. Descriptive data of the number of muscle synergies, global VAF each group.

	Unant Mean (SD)	Ant Mean (SD)	p-Value
The number of muscle synergies		4.15(0.62)	
Global VAF (%)			
With three synergies	84.93 (1.54)	85.28 (1.51)	0.153
With four synergies	90.03 (0.94)	90.53 (1.19)	0.139
With five synergies	92.4 (0.80)	92.71 (0.89)	0.192
Local VAFs with four synergies (%)			

Table 1. *Cont.*

	Unant Mean (SD)	Ant Mean (SD)	*p*-Value
GM	94.83 (3.23)	89.37 (4.79)	0.105
RF	89.32 (4.23)	88.18 (3.37)	0.661
VM	93.75 (3.10)	91.41 (2.87)	0.327
VL	89.73 (4.16)	88.54 (4.97)	0.504
MG	93.26 (2.01)	95.14 (2.56)	0.411
LG	92.19 (6.35)	96.13 (3.11)	0.226
SL	91.75 (3.01)	92.36 (1.93)	0.723
TA	92.46 (4.58)	92.15 (3.53)	0.869
PL	94.06 (5.03)	90.11 (2.82)	0.202

Note: Unant: Unanticipated group; Ant: Anticipated group; SD: standard deviation.

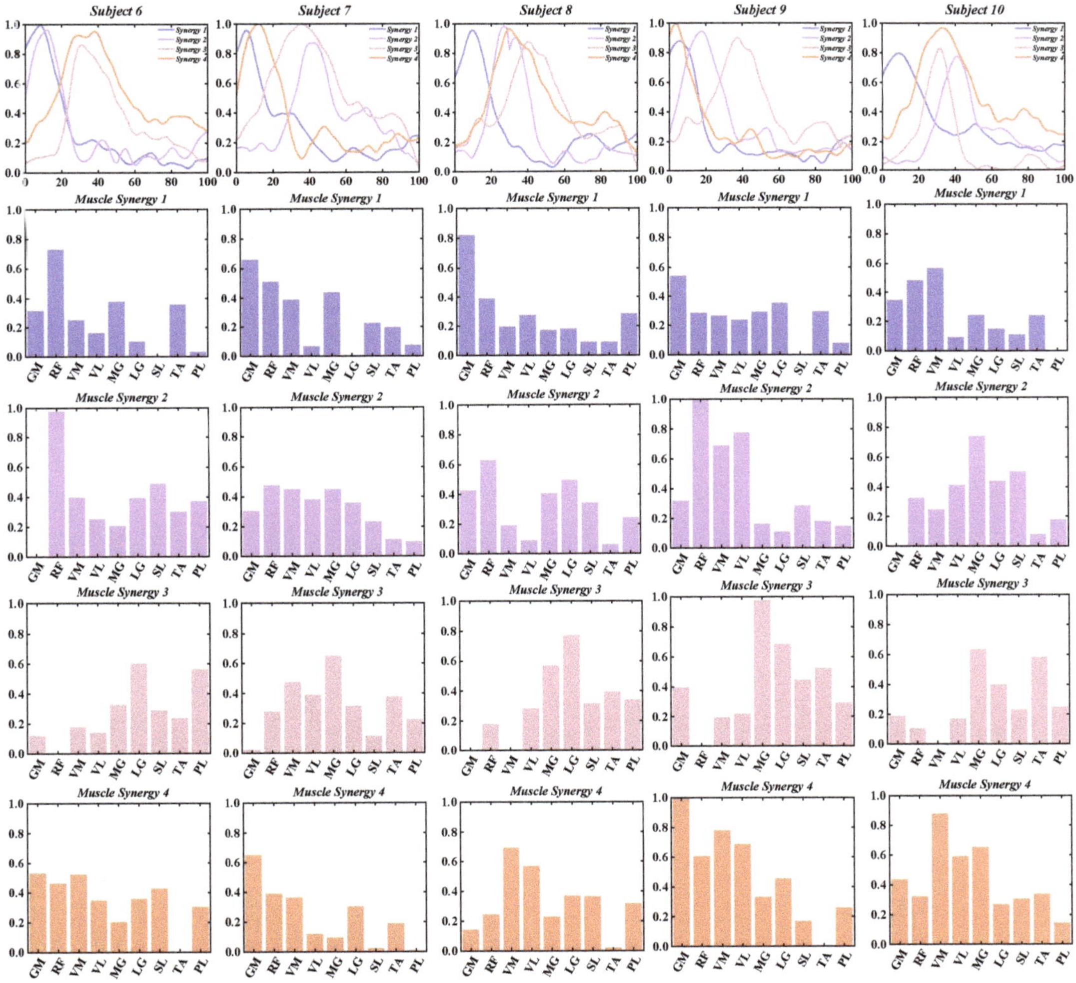

Figure 6. Detailed Presentation of activation coefficient results and synergy vectors for Subjects 6–10 in the Anticipated Landing Group.

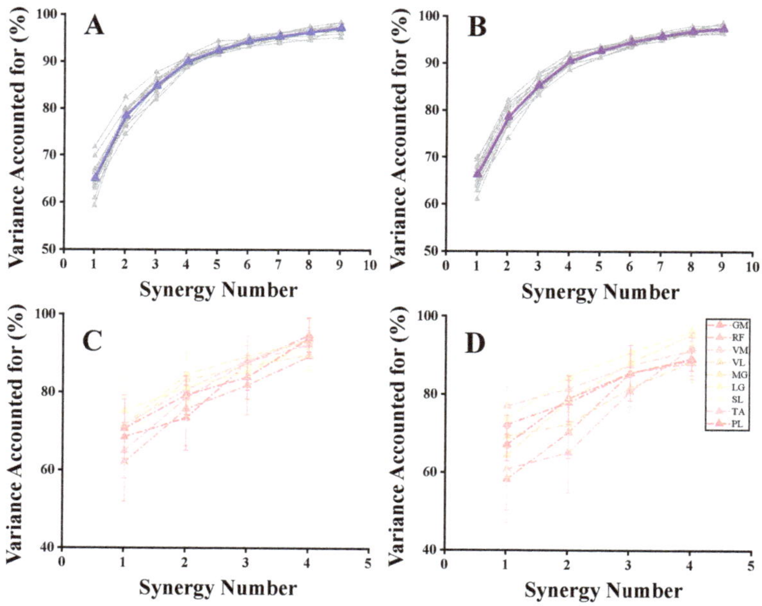

Figure 7. The global and local VAF correspond to each synergy in the Unanticipated and Anticipated groups. (**A**) The global VAF of the unanticipated group; (**B**) The global VAF of the anticipated group; (**C**) The local VAF of the unanticipated group; (**D**) The local VAF of the anticipated group. The gray line illustrates the VAF of each subject, the blue line illustrates the VAF of the unanticipated group and the purple line illustrates the VAF of the anticipated group.

3.2. Similarity of Muscle Synergies

Using the Silhouette coefficient to assess the clustering quality, we identified that the clustering effect was maximized when i = 4. This indicates that the optimal number of clusters was four, leading us to use three clustering centers for the reference muscle synergies (Figure 8).

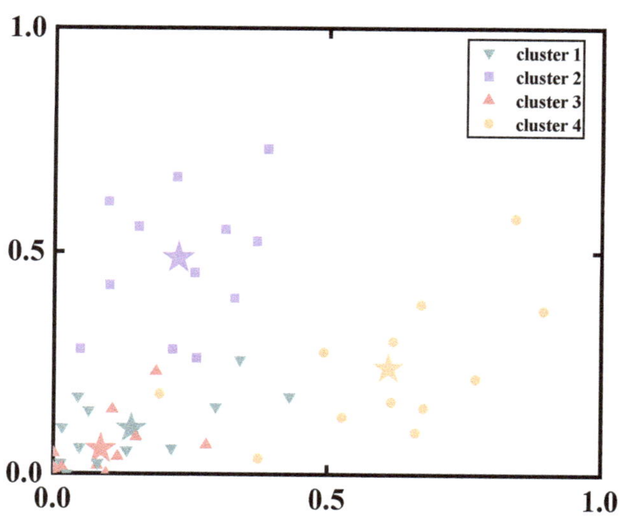

Figure 8. Visualizations of K-means clustering results for all synergy vectors in the Anticipated group. The asterisk represents the cluster center.

By using the four cluster centroids as a basis for synergy comparison and applying Pearson correlation coefficients to organize the synergy patterns of individual participants (Figures 9 and 10), we observed that the similarity of reference synergies in the anticipated landing group was significantly higher than in the unanticipated landing group.

Figure 9. Correlation coefficients between synergy vectors and reference synergies for the 20 participants during the unanticipated landing tasks. W: synergy vectors. RS: Reference Synergy.

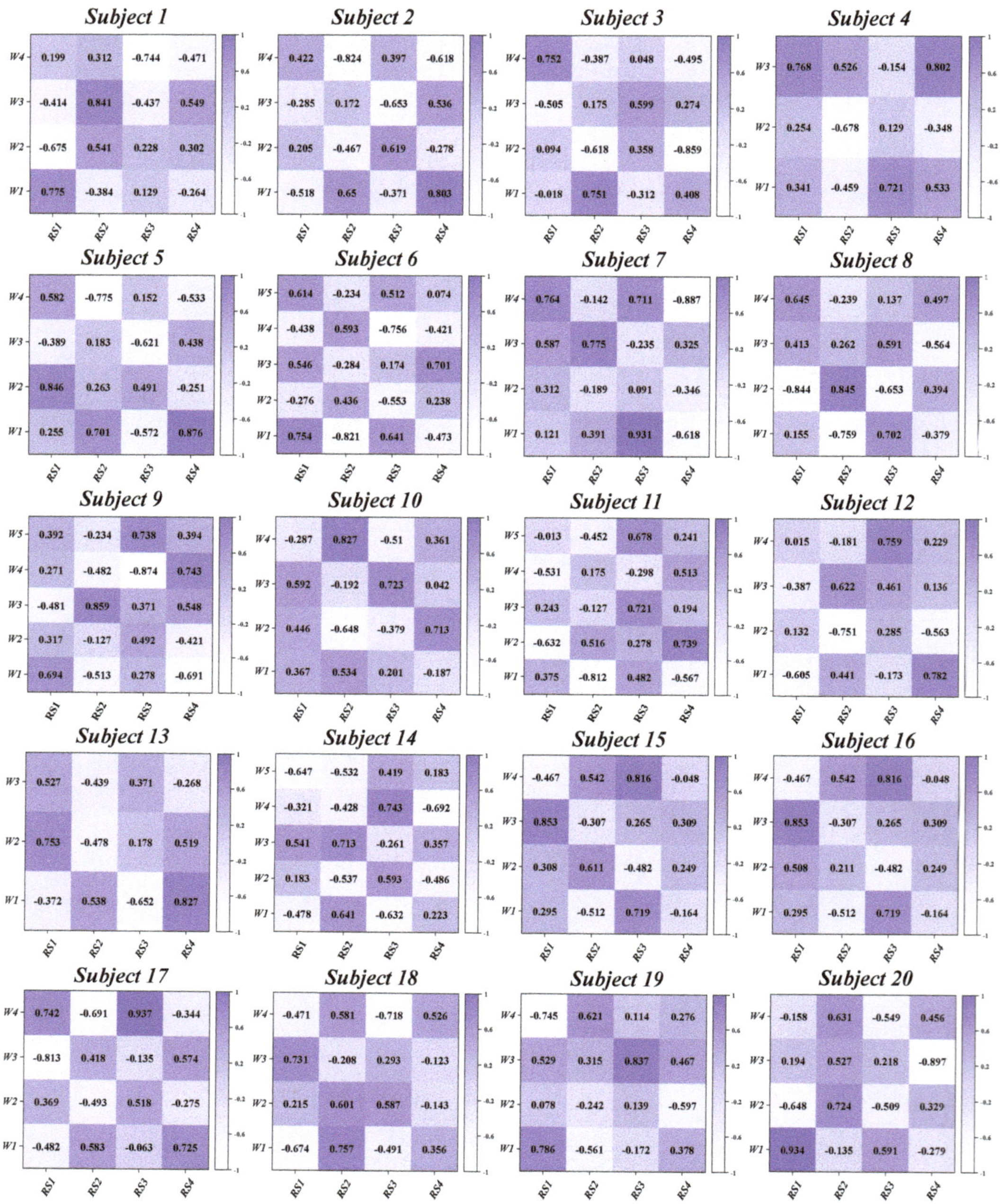

Figure 10. Correlation coefficients between synergy vectors and reference synergies for the 20 participants during the anticipated landing tasks. W: synergy vectors. RS: Reference Synergy.

3.3. The Functional Interpretation of Muscle Synergies

Based on the muscle synergy analysis, we defined muscle weight exceeding 0.3 as indicative of activation. The results revealed distinct activation patterns for different muscle synergies during the landing process (Figure 11A).

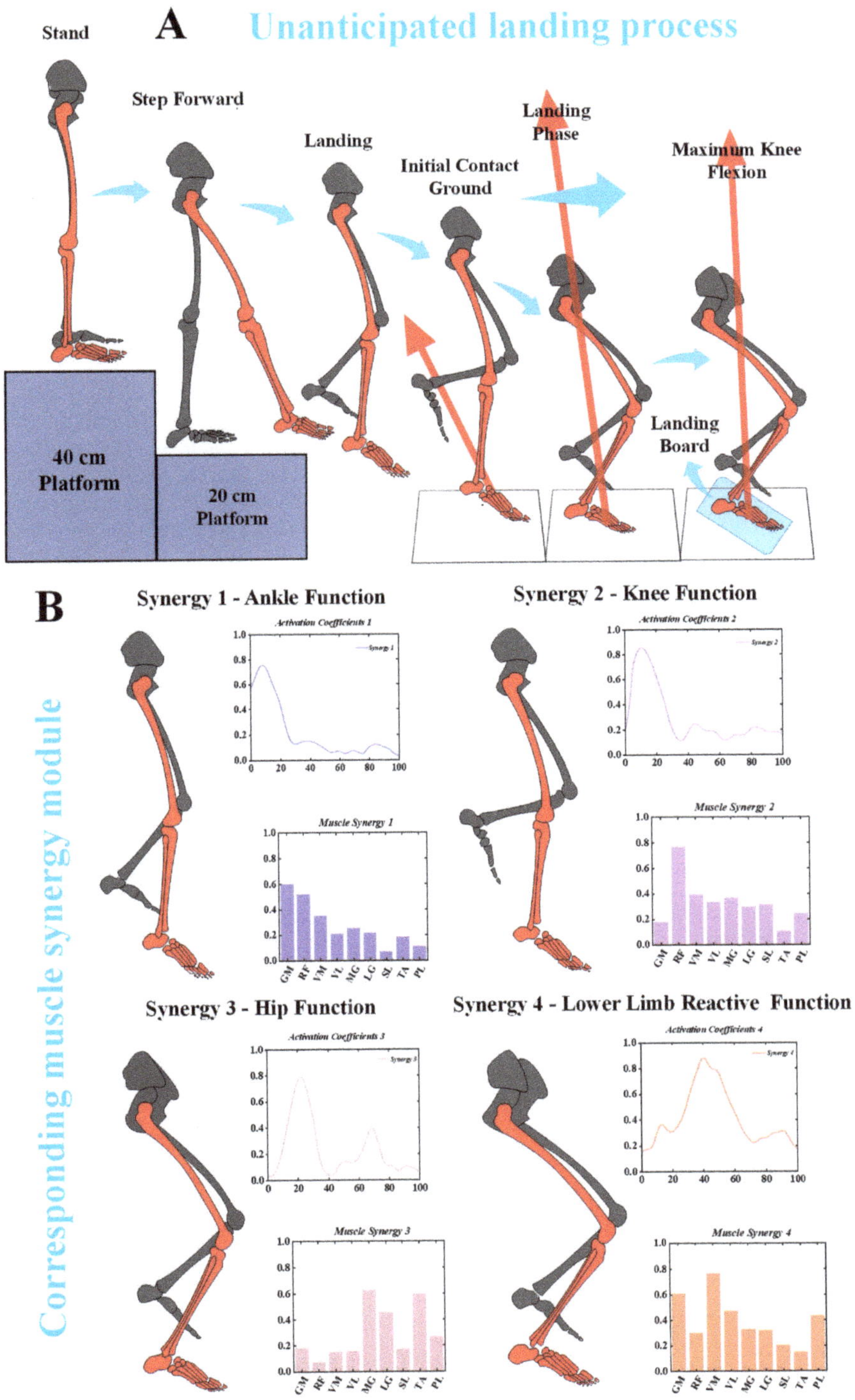

Figure 11. The specific landing biomechanical test and corresponding muscle synergy module. (**A**) Unanticipated landing biomechanical test. (**B**) Corresponding muscle synergy module.

In Synergy 1, the activation levels of the gluteus maximus (GM), rectus femoris (RF), and vastus medialis (VM) were relatively high, suggesting that this synergy is closely associated with hip joint function during the landing phase. Synergy 2 displayed high activation levels in the RF, vastus medialis (VM), vastus lateralis (VL), lateral gastrocnemius (LG), medial gastrocnemius (MG), and soleus (SL). This pattern is likely related to the stabilization of the knee joint during landing. Synergy 3 primarily involved the MG, LG, and tibialis anterior (TA), with significant activation occurring during the initial 0–20% of the landing phase. This indicates the critical role of these muscles in controlling the ankle during the initial contact phase. Finally, Synergy 4 showed activation in multiple muscle groups, including the GM, VM, VL, LG, MG, and PL. This pattern highlights that CAI patients rely on the coordinated effort of several muscle groups to maintain stability during the latter part of the landing phase (Figure 11B).

3.4. Specific Characteristics of Muscle Synergy

Based on the results from Muscle Synergy Vectors (Figure 12), in Synergy 1, we observed that the relative weights of most specific muscles did not show significant differences between the Unanticipated and Anticipated groups. The only notable difference was that the relative weight of the GM in the Unanticipated group was significantly higher than that in the anticipated group ($p = 0.041$). In Synergy 2, compared to the Anticipated group, the VL in the Unanticipated group was significantly larger, with a notable difference between the two groups ($p = 0.043$). In Synergy 3, the MG in the Anticipated group was significantly larger than that in the Unanticipated group ($p < 0.05$), while the TA in the Unanticipated group was significantly larger than that in the Anticipated group ($p < 0.05$). In Synergy 4, the GM ($p < 0.05$) and the VM ($p < 0.001$) in the Unanticipated group were significantly greater than those in the Anticipated group. Conversely, the MG ($p = 0.044$) and the LG ($p = 0.037$) in the Anticipated group were significantly greater than those in the Unanticipated group (Table 2).

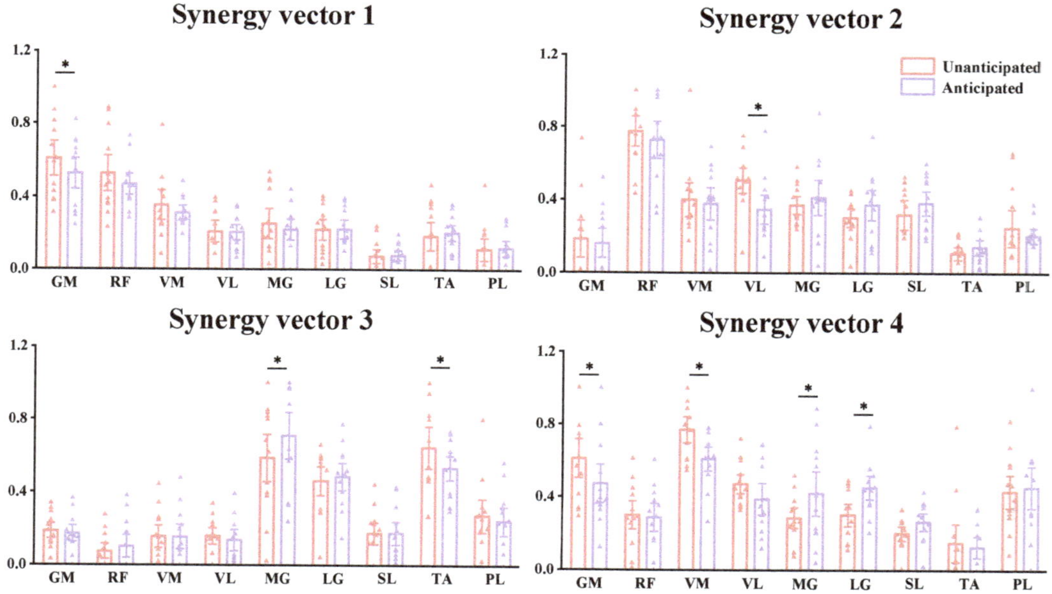

Figure 12. Muscle synergy vectors extracted from the Unanticipated and Anticipated groups. * significant difference with $p < 0.05$.

Table 2. Comparison of muscle synergy vectors during landing between the Unanticipated group and Anticipated group.

Synergy Vectors Mean (SD)	Muscle Synergy 1			Muscle Synergy 2			Muscle Synergy 3			Muscle Synergy 4		
	Unant	Ant	*p*-Value	Unant	Ant	*p*-Value	Unant	Ant	*p*-Value	Unant	Ant	*p*-Value
IGM	0.61 (0.22)	0.53 (0.19)	0.041	0.20 (0.12)	0.18 (0.07)	0.679	0.19 (0.09)	0.17 (0.10)	0.733	0.61 (0.25)	0.47 (0.24)	0.038
IRF	0.53 (0.22)	0.47 (0.13)	0.287	0.77 (0.19)	0.73 (0.23)	0.432	0.08 (0.10)	0.10 (0.14)	0.515	0.30 (0.18)	0.28 (0.17)	0.688
IVM	0.35 (0.18)	0.31 (0.10)	0.376	0.55 (0.25)	0.38 (0.20)	0.040	0.16 (0.14)	0.16 (0.15)	0.996	0.77 (0.18)	0.61 (0.15)	0.001
IVL	0.21 (0.13)	0.20 (0.10)	0.953	0.50 (0.16)	0.35 (0.18)	0.014	0.16 (0.10)	0.14 (0.13)	0.583	0.47 (0.12)	0.39 (0.21)	0.140
IMG	0.25 (0.18)	0.22 (0.13)	0.369	0.37 (0.11)	0.41 (0.22)	0.517	0.59 (0.30)	0.71 (0.17)	0.039	0.29 (0.13)	0.42 (0.22)	0.044
ILG	0.22 (0.12)	0.23 (0.11)	0.939	0.30 (0.14)	0.37 (0.19)	0.285	0.46 (0.19)	0.48 (0.17)	0.592	0.30 (0.12)	0.45 (0.16)	0.037
ISL	0.07 (0.08)	0.08 (0.06)	0.808	0.32 (0.19)	0.38 (0.15)	0.397	0.17 (0.04)	0.18 (0.04)	0.974	0.20 (0.10)	0.27 (0.12)	0.123
ITA	0.19 (0.17)	0.21 (0.09)	0.712	0.11 (0.07)	0.14 (0.09)	0.348	0.67 (0.27)	0.51 (0.18)	0.040	0.15 (0.24)	0.13 (0.12)	0.731
IPL	0.11 (0.14)	0.12 (0.10)	0.856	0.33 (0.06)	0.20 (0.10)	0.049	0.27 (0.06)	0.24 (0.05)	0.697	0.43 (0.21)	0.46 (0.26)	0.699

Note: Unant: Unanticipated group; Ant: Anticipated group; SD: standard deviation.

4. Discussion

Our study aimed to explore and compare the neuromuscular control strategies of individuals with chronic ankle instability (CAI) during anticipated and unanticipated landing tasks using non-negative matrix factorization (NNMF) combined with muscle synergy extraction techniques. This study offers the first comprehensive assessment of how CAI patients organize their neuromuscular responses and strategies during unanticipated landings. Our study reveals several key findings: (1) During landing tasks, CAI patients exhibit significant activation of proximal muscles, particularly the gluteus maximus and rectus femoris, demonstrating distinct muscle synergy patterns. (2) In unanticipated landing tasks, CAI patients show an active neuromuscular response with notable adjustments in their ankle and knee control strategies. (3) During unanticipated landings, CAI patients display specific muscle synergy patterns, reflecting how they adapt to unforeseen landing scenarios through various muscle coordination strategies to maintain stability.

Our study reveals that during unanticipated landings, the weight proportions of the gluteus maximus, rectus femoris, and vastus medialis in muscle synergy 1 are significantly higher in individuals with CAI compared to anticipated landings. This finding indicates that proximal joint muscles play a crucial role in managing unanticipated landing tasks in CAI patients [31]. The notable activation of these proximal muscles in such sudden landing situations reflects their central role in maintaining body stability and control [32]. Specifically, our research shows that CAI patients exhibit higher levels of proximal muscle activation during unanticipated landings compared to anticipated landings. This suggests that CAI patients rely on the gluteus maximus, rectus femoris, and vastus medialis to enhance stability and compensate for the compromised ankle function when faced with sudden landing impacts. This phenomenon is consistent with previous studies, which have demonstrated significant biomechanical changes in the proximal hip joint of CAI patients during landings. Ankle joint impairment compels CAI patients to activate hip joint adjustment mechanisms during unanticipated landings to compensate for the functional deficits of the ankle [33–35].

Another significant finding of our study is that in synergy 2, the weight of the gluteus maximus and vastus lateralis are notably higher in CAI patients during unanticipated landings compared to anticipated landings, whereas during anticipated landings, the weight of the medial and lateral gastrocnemius is higher. This suggests that CAI patients tend to rely more on proximal knee muscles to enhance body stability during unanticipated landings. This compensatory mechanism likely arises because ankle dysfunction hinders the joint from providing sufficient support during landings, shifting more of the stability task to the proximal muscles around the knee. This compensatory strategy highlights the critical role of proximal muscles in maintaining balance, especially when CAI patients are confronted with sudden, unanticipated landings. The activation of these proximal muscles not only provides support for dynamic balance but also helps absorb the impact forces during landing. Beck et al.'s research further supports this notion, suggesting that activating proximal knee muscles (such as knee flexors) consumes more energy compared to distal muscles (such as ankle plantar flexors), indicating that proximal muscles play a more significant role in absorbing impact and maintaining balance [36]. Furthermore, the importance of the knee in managing landing impacts and maintaining dynamic balance underscores its central role in the compensatory strategies of CAI patients [37]. When ankle function is compromised, the knee becomes a substitute source of power by mobilizing surrounding proximal muscle groups, effectively distributing external forces during landings. This reduces the load on the ankle joint, minimizing the risk of further injury [38,39].

In Synergy 3, the weight of the tibialis anterior (TA) in CAI patients during unanticipated landings is significantly higher compared to anticipated landings. This finding suggests that CAI patients compensate for their lack of ankle stability by increasing the activation of ankle dorsiflexion during unanticipated landings. The enhanced activation of the TA helps stabilize the ankle in a tighter, more secure position, thereby improving overall landing stability. Particularly in response to unanticipated landings, the activation of the TA plays a critical role in controlling the center of mass and preventing imbalance. Gribble's research on the motor control strategies of CAI patients found that they increase co-activation across major lower limb joints, especially in unstable or dynamic environments, which aligns with our finding regarding the role of the TA during unanticipated landings [40]. The study also highlighted the importance of early TA activation during the initial phase of landing, particularly in terms of ankle co-activation and overall stability. By activating the TA in advance, CAI patients can enhance the support of the ankle joint, better managing unpredictable landing conditions and maintaining dynamic balance. This early response mechanism of the TA not only helps prevent excessive ankle inversion and eversion but also reduces the risk of injury. DeMers' musculoskeletal modeling study further highlighted that increased contraction of both ankle invertors and evertors enhances overall ankle stability, thereby reducing the risk of potential injuries [41]. By strengthening the use of ankle dorsiflexion and the co-activation mechanism, CAI patients can effectively compensate for their lack of ankle stability, particularly during unanticipated landings, and prevent further damage. Additionally, Kim emphasized the critical role of the TA in compensating for ankle sprain-related deficiencies [42]. Effective activation of the TA enables CAI patients to better control ankle posture and enhance overall stability during dynamic activities. By increasing TA activation, CAI patients can quickly lock the ankle joint during landing, reducing the potential risks associated with ankle instability. This further validates the key role of the TA in maintaining dynamic balance and preventing ankle injuries [43].

Analysis of synergy 4 reveals that the muscle synergy patterns of individuals with CAI during unanticipated landings differ from those observed during anticipated landings. During unanticipated landings, the activation weight of the gluteus maximus and vastus lateralis is significantly higher compared to anticipated landings. This finding indicates that CAI patients rely more on proximal muscles to enhance body stability when faced with sudden landing impacts [34,44]. This compensatory mechanism underscores the critical role of proximal muscles in maintaining stability. Additionally, we observed that during anticipated landings, the activation weights of the medial and lateral gastrocnemius are notably higher than during unanticipated landings. This suggests that CAI patients rely more on the posterior muscle groups of the ankle joint to control and absorb landing impacts during anticipated tasks. The different muscle activation patterns reflect how CAI patients adjust their muscle control strategies to meet the challenges presented by different types of landing tasks. This finding is supported by several researchers. For instance, DeJong et al. noted that CAI patients exhibit significant compensatory muscle activation in proximal joints (such as the knee and hip), indicating that impaired ankle function prompts patients to enhance proximal muscle activity to compensate for ankle deficiencies [45]. This compensatory mechanism is particularly evident during unanticipated landing tasks, where patients rely more on proximal muscles to make up for ankle deficiencies. Kim et al. observed increased activation of knee and hip muscles (such as the vastus lateralis, vastus medialis, gluteus maximus, and gluteus medius) during the transition phase of landing and cutting tasks in CAI patients, which further supports our findings [46]. Rios et al. found that CAI patients demonstrate notably increased activation in proximal muscles during single-leg stance tasks, in contrast to the healthy group, suggesting that CAI patients improve postural control and reduce ankle instability by increasing the activation of proximal muscles [47,48]. In contrast, during anticipated landing tasks, the activation weight of the medial and lateral gastrocnemius muscles is elevated. This phenomenon is consistent with the findings of Gehring et al., who highlighted that in CAI patients, the posterior muscle groups of the ankle (such as the gastrocnemius) play a crucial role in controlling and absorbing landing impacts during anticipated shocks [33,49]. This muscle activation pattern helps patients effectively disperse external forces and reduce the load on the ankle joint during anticipated landings [50]. These findings stress the significance of focusing on proximal muscle function when designing rehabilitation and prevention strategies for CAI patients, while also revealing differences in ankle muscle activation patterns across various landing tasks. Such insights provide a solid foundation for further research and clinical interventions aimed at helping CAI patients improve their motor control strategies, enhance stability, and reduce the risk of injury.

Despite providing valuable insights into the muscle synergy patterns of CAI patients across different landing tasks, this study has several limitations. Notably, the sample size is relatively small, encompassing just 20 CAI patients. A smaller sample size may affect the determination of the optimal number of muscle synergies and could introduce errors in the variance accounted for (VAF), impacting the accuracy of the results. Additionally, a limited sample size may introduce bias during the K-means clustering process, affecting the reliability of the findings. Secondly, this study compared muscle synergy patterns only in CAI patients during anticipated and unanticipated landings, without comparing these patterns to a healthy control group. This limitation restricts our understanding of the differences in muscle synergy between CAI patients and healthy individuals and does not reveal the muscle activation patterns of the healthy group during the same tasks. This comparison is crucial for a deeper understanding of the pathological mechanisms of CAI and the development of effective rehabilitation strategies. Thirdly, the study lacks the recording and interpretation of kinematic and dynamic data. This gap hinders a comprehensive analysis of muscle synergies and their clinical implications. The lack of kinematic and dynamic data limits the exploration of how muscle activation relates to movement performance, thereby affecting our overall understanding of motor control in CAI patients during landings. Finally, the current research primarily focuses on muscle synergy patterns during the landing process, which may be somewhat limited for studies on ankle injury patients. Future research should extend to other movement domains, such as cutting, walking, running, and jumping, to comprehensively assess muscle control strategies under various movement conditions. This approach will help in understanding the characteristics of muscle synergy across different movement patterns.

5. Conclusions

This study investigated the changes in muscle synergy patterns in patients with CAI during anticipated and unanticipated landing tasks and the influence of these changes on neuromuscular control strategies. The findings demonstrated that CAI patients exhibit distinct muscle synergy patterns across the two landing conditions, which prompts the adoption of different neuromuscular control strategies to cope with the varying demands of each task. In unanticipated landings, CAI patients rely more heavily on the activation of proximal muscles, such as the gluteus maximus and vastus lateralis. This compensatory activation emphasizes their attempt to bolster hip and knee stability, compensating for the impaired function at the ankle joint. In contrast, during anticipated landings, CAI patients predominantly engage the posterior ankle muscles, particularly the medial and lateral gastrocnemius, to better manage and absorb the expected landing impact. These differences in muscle activation patterns suggest that CAI patients adapt their neuromuscular control strategies in response to the specific challenges posed by different landing tasks. This nuanced understanding of task-dependent muscle synergy adaptations may provide valuable insights for rehabilitation approaches aimed at improving functional stability in CAI patients.

Author Contributions: Z.Z., D.X. and Y.G. conceived and designed the study; Z.Z., D.X. and Y.G. recruited and screened subjects for the experiments; Z.Z., H.Z. and Y.G. performed the experimental data acquisition and recording; D.X. and H.Z. validated the OpenSim musculoskeletal modeling electromyography data and EMG data; M.W. and T.J. performed the data trend analysis; Z.Z. and H.Z. performed the data visualization; Z.Z. and Z.Z. performed the original draft writing; Z.Z., J.S.B. and T.J. performed the data visualization; M.W. and J.S.B. performed data trend analysis; Z.Z. and H.Z. performed data visualization; Z.Z. wrote the original draft; T.J., M.W. and Y.G. reviewed and edited the writing; D.X., H.Z. and Y.G. supervised the study; and Y.G. secured funding. All authors have read and agreed to the published version of the manuscript.

Funding: This research was sponsored by Zhejiang Provincial Natural Science Foundation of China for Distinguished Young Scholars (LR22A020002), Zhejiang Provincial Key Research and Development Program of China (2023C03197), Ningbo Key R&D Program (2022Z196), Zhejiang Province Exploring Public Welfare Projects (LTGY23H040003), Ningbo Natural Science Foundation (20221JCGY010532, 20221JCGY010607), Public Welfare Science and Technology Project of Ningbo, China (2021S134), and Zhejiang Rehabilitation Medical Association Scientific Research Special Fund (RAGH20240622).

Institutional Review Board Statement: The study was conducted by the Declaration of Helsinki and approved by the Ethics Committee of Ningbo University approved this study (protocol code: TY2024001).

Informed Consent Statement: Informed consent was obtained from the subject involved in the study.

Data Availability Statement: Data are available on request due to the restriction of ethics.

Conflicts of Interest: The authors declare no conflicts of interest.

References

1. Doherty, C.; Delahunt, E.; Caulfield, B.; Hertel, J.; Ryan, J.; Bleakley, C. The incidence and prevalence of ankle sprain injury: A systematic review and meta-analysis of prospective epidemiological studies. *Sports Med.* **2014**, *44*, 123–140. [CrossRef]
2. Fong, D.T.-P.; Hong, Y.; Chan, L.-K.; Yung, P.S.-H.; Chan, K.-M. A systematic review on ankle injury and ankle sprain in sports. *Sports Med.* **2007**, *37*, 73–94. [CrossRef] [PubMed]
3. Wikstrom, E.A.; Hubbard-Turner, T.; McKeon, P.O. Understanding and treating lateral ankle sprains and their consequences: A constraints-based approach. *Sports Med.* **2013**, *43*, 385–393. [CrossRef] [PubMed]
4. Liew, B.X.; Sullivan, L.; Morris, S.; Netto, K. Mechanical work performed by distal foot-ankle and proximal knee-hip segments during anticipated and unanticipated cutting. *J. Biomech.* **2020**, *106*, 109839. [CrossRef] [PubMed]
5. van der Merwe, C.; Shultz, S.P.; Colborne, G.R.; Hébert-Losier, K.; Fink, P.W. The coordination patterns of the foot segments in relation to lateral ankle sprain injury mechanism during unanticipated changes of direction. *Foot* **2020**, *45*, 101745. [CrossRef]
6. Kim, H.; Palmieri-Smith, R.; Kipp, K. Muscle force contributions to ankle joint contact forces during an unanticipated cutting task in people with chronic ankle instability. *J. Biomech.* **2021**, *124*, 110566. [CrossRef]
7. Fong, C.-M.; Blackburn, J.T.; Norcross, M.F.; McGrath, M.; Padua, D.A. Ankle-dorsiflexion range of motion and landing biomechanics. *J. Athl. Train.* **2011**, *46*, 5–10. [CrossRef]
8. Jang, J.; Wikstrom, E.A. Ankle joint contact force profiles differ between those with and without chronic ankle instability during walking. *Gait Posture* **2023**, *100*, 1–7. [CrossRef]

9. Zhang, S.-N.; Bates, B.T.; Dufek, J.S. Contributions of lower extremity joints to energy dissipation during landings. *Med. Sci. Sports Exerc.* **2000**, *32*, 812–819. [CrossRef]
10. Zhou, Z.; Zhou, H.; Jie, T.; Xu, D.; Teo, E.-C.; Wang, M.; Gu, Y. Analysis of stress response distribution in patients with lateral ankle ligament injuries: A study of neural control strategies utilizing predictive computing models. *Front. Physiol.* **2024**, *15*, 1438194. [CrossRef]
11. Xu, D.; Zhou, H.; Quan, W.; Gusztav, F.; Baker, J.S.; Gu, Y. Adaptive neuro-fuzzy inference system model driven by the non-negative matrix factorization-extracted muscle synergy patterns to estimate lower limb joint movements. *Comput. Methods Programs Biomed.* **2023**, *242*, 107848. [CrossRef] [PubMed]
12. Lee, D.D.; Seung, H.S. Learning the parts of objects by non-negative matrix factorization. *Nature* **1999**, *401*, 788–791. [CrossRef]
13. Rabbi, M.F.; Pizzolato, C.; Lloyd, D.G.; Carty, C.P.; Devaprakash, D.; Diamond, L.E. Non-negative matrix factorisation is the most appropriate method for extraction of muscle synergies in walking and running. *Sci. Rep.* **2020**, *10*, 8266. [CrossRef] [PubMed]
14. Ambrosini, E.; Parati, M.; Peri, E.; De Marchis, C.; Nava, C.; Pedrocchi, A.; Ferriero, G.; Ferrante, S. Changes in leg cycling muscle synergies after training augmented by functional electrical stimulation in subacute stroke survivors: A pilot study. *J. Neuroeng. Rehabil.* **2020**, *17*, 35. [CrossRef] [PubMed]
15. McVey, E.D.; Palmieri, R.M.; Docherty, C.L.; Zinder, S.M.; Ingersoll, C.D. Arthrogenic muscle inhibition in the leg muscles of subjects exhibiting functional ankle instability. *Foot Ankle Int.* **2005**, *26*, 1055–1061. [CrossRef] [PubMed]
16. Chvatal, S.A.; Ting, L.H. Common muscle synergies for balance and walking. *Front. Comput. Neurosc.* **2013**, *7*, 48. [CrossRef]
17. Jie, T.; Xu, D.; Zhang, Z.; Teo, E.-C.; Baker, J.S.; Zhou, H.; Gu, Y. Structural and Organizational Strategies of Locomotor Modules during Landing in Patients with Chronic Ankle Instability. *Bioengineering* **2024**, *11*, 518. [CrossRef]
18. Ghislieri, M.; Labanca, L.; Mosca, M.; Bragonzoni, L.; Knaflitz, M.; Benedetti, M.G.; Agostini, V. Balance and Muscle Synergies During a Single-Limb Stance Task in Individuals With Chronic Ankle Instability. *IEEE Trans. Neural Syst. Rehabil. Eng.* **2023**, *31*, 4367–4375. [CrossRef]
19. Hermens, H.J.; Freriks, B.; Merletti, R.; Stegeman, D.; Blok, J.; Rau, G.; Disselhorst-Klug, C.; Hägg, G. European recommendations for surface electromyography. *IEEE Trans. Neural Syst. Rehabil. Eng.* **1999**, *8*, 13–54.
20. Herzog, W.; Sokolosky, J.; Zhang, Y.; Guimarães, A. EMG-force relation in dynamically contracting cat plantaris muscle. *J. Electromyogr. Kines* **1998**, *8*, 147–155. [CrossRef]
21. Lee, D.; Seung, H.S. Algorithms for non-negative matrix factorization. In *Advances in Neural Information Processing Systems*; MIT Press: Cambridge, MA, USA, 2000; p. 13.
22. Sedaghat-Nejad, E.; Mousavi, S.J.; Hadizadeh, M.; Narimani, R.; Khalaf, K.; Campbell-Kyureghyan, N.; Parnianpour, M. Is there a reliable and invariant set of muscle synergy during isometric biaxial trunk exertion in the sagittal and transverse planes by healthy subjects? *J. Biomech.* **2015**, *48*, 3234–3241. [CrossRef] [PubMed]
23. Boccia, G.; Zoppirolli, C.; Bortolan, L.; Schena, F.; Pellegrini, B. Shared and task-specific muscle synergies of Nordic walking and conventional walking. *Scand. J. Med. Sci. Spor.* **2018**, *28*, 905–918. [CrossRef] [PubMed]
24. Li, X.; Zeng, H.; Li, Y.; Song, A. Quantitative Assessment via Multi-Domain Fusion of Muscle Synergy Associated with Upper-Limb Motor Function for Stroke Rehabilitation. *IEEE Trans. BioMed. Eng.* **2024**, *71*, 1430–1441. [CrossRef] [PubMed]
25. Krishna, K.; Murty, M.N.; Cybernetics, P.B. Genetic K-means algorithm. *IEEE Trans. Syst. Man Cybern. Part B* **1999**, *29*, 433–439. [CrossRef]
26. Modha, D.S.; Spangler, W.S. Feature weighting in k-means clustering. *Mach. Learn.* **2003**, *52*, 217–237. [CrossRef]
27. Kim, T.K. T test as a parametric statistic. *Korean J. Anesthesiol.* **2015**, *68*, 540–546. [CrossRef]
28. Xu, D.; Lu, J.; Baker, J.S.; Fekete, G.; Gu, Y. Temporal kinematic and kinetics differences throughout different landing ways following volleyball spike shots. *Proc. Inst. Mech. Eng. Part P J. Sports Eng. Technol.* **2022**, *236*, 200–208. [CrossRef]
29. Pataky, T.C. One-dimensional statistical parametric mapping in Python. *Comput. Methods Biomech. Biomed. Eng.* **2012**, *15*, 295–301. [CrossRef]
30. Xu, D.; Zhou, H.; Quan, W.; Jiang, X.; Liang, M.; Li, S.; Ugbolue, U.C.; Baker, J.S.; Gusztav, F.; Ma, X.; et al. A new method proposed for realizing human gait pattern recognition: Inspirations for the application of sports and clinical gait analysis. *Gait Posture* **2024**, *107*, 293–305. [CrossRef]
31. Zhou, Z.F.; Xu, D.T.; Zhou, H.Y.; Chon, T.E.; Baker, J.S.; Gu, Y.D. Distinct Motion Control Strategy during Unanticipated Landing: Transitioning from Copers to Chronic Ankle Instability. *J. Biomim. Biomater. Biomed. Eng.* **2024**, *65*, 15–32. [CrossRef]
32. Zhou, H.; Ugbolue, U.C. Biomechanical analysis of lower limbs based on unstable condition sports footwear: A systematic review. *Phys. Act. Health* **2024**, *8*, 93–104. [CrossRef]
33. Gehring, D.; Wissler, S.; Mornieux, G.; Gollhofer, A. How to sprain your ankle–a biomechanical case report of an inversion trauma. *J. Biomech.* **2013**, *46*, 175–178. [CrossRef] [PubMed]
34. DeJong, A.F.; Koldenhoven, R.M.; Hertel, J. Hip biomechanical alterations during walking in chronic ankle instability patients: A cross-correlation analysis. *Sports Med.* **2022**, *21*, 460–471. [CrossRef]
35. Li, F.; Xu, D.; Zhou, H.; Kovács, B.; Liang, M.J.I. The effect of heel height on the Achilles tendon and muscle activity in Latin dancers during a special-landing task. *Int. J. Biomed. Eng. Technol.* **2024**, *44*, 303–323. [CrossRef]
36. Beck, J.J.; VandenBerg, C.; Cruz, A.I.; Ellis Jr, H.B. Low energy, lateral ankle injuries in pediatric and adolescent patients: A systematic review of ankle sprains and nondisplaced distal fibula fractures. *J. Pediater. Orthoped.* **2020**, *40*, 283–287. [CrossRef]

37. Han, S.; Lee, H.; Oh, M.; Hopkins, J.T. Lower Extremity Energy Dissipation and Generation During Jump Landing and Cutting in Patients with Chronic Ankle Instability. *J. Athl. Train.* **2023**, *58*, 912–919. [CrossRef]
38. Mineta, S.; Fukano, M.; Hirose, N. Less impact absorption at the ankle joint is related to the single-leg landing stability deficit in patients with chronic ankle instability. *J. Biomech.* **2023**, *149*, 111509. [CrossRef]
39. Xu, D.; Zhou, H.; Quan, W.; Gusztav, F.; Wang, M.; Baker, J.S.; Gu, Y. Accurately and effectively predict the ACL force: Utilizing biomechanical landing pattern before and after-fatigue. *Comput. Methods Programs Biomed.* **2023**, *241*, 107761. [CrossRef]
40. Gribble, P.A.; Bleakley, C.M.; Caulfield, B.M.; Docherty, C.L.; Fourchet, F.; Fong, D.T.-P.; Hertel, J.; Hiller, C.E.; Kaminski, T.W.; McKeon, P.O.; et al. Evidence review for the 2016 International Ankle Consortium consensus statement on the prevalence, impact and long-term consequences of lateral ankle sprains. *Br. J. Sports Med.* **2016**, *50*, 1496–1505. [CrossRef]
41. DeMers, M.S.; Pal, S.; Delp, S.L. Changes in tibiofemoral forces due to variations in muscle activity during walking. *J. Orthop. Res.* **2014**, *32*, 769–776. [CrossRef]
42. Kim, H.; Palmieri-Smith, R.; Kipp, K. Muscle synergies in people with chronic ankle instability during anticipated and unanticipated landing-cutting tasks. *J. Athl. Train.* **2023**, *58*, 143–152. [CrossRef] [PubMed]
43. Kim, H.; Palmieri-Smith, R.; Kipp, K. Time-frequency analysis of muscle activation patterns in people with chronic ankle instability during Landing and cutting tasks. *Gait Posture* **2020**, *82*, 203–208. [CrossRef] [PubMed]
44. DeJong, A.F.; Mangum, L.C.; Hertel, J. Gluteus medius activity during gait is altered in individuals with chronic ankle instability: An ultrasound imaging study. *Gait Posture* **2019**, *71*, 7–13. [CrossRef]
45. DeJong, A.F.; Koldenhoven, R.M.; Hart, J.M.; Hertel, J. Gluteus medius dysfunction in females with chronic ankle instability is consistent at different walking speeds. *Clin. Biomech.* **2020**, *73*, 140–148. [CrossRef]
46. Choi, W.J.; Lee, J.W.; Han, S.H.; Kim, B.S.; Lee, S.K. Chronic lateral ankle instability: The effect of intra-articular lesions on clinical outcome. *Am. J. Sports Med.* **2008**, *36*, 2167–2172. [CrossRef]
47. dos Santos, M.J.; Gorges, A.L.; Rios, J. Individuals with chronic ankle instability exhibit decreased postural sway while kicking in a single-leg stance. *Gait Posture* **2014**, *40*, 231–236. [CrossRef]
48. Yu, P.; Fernandez, J. Alterations in Lower Limb Biomechanical Characteristics During the Cutting Manoeuvre in Chronic Ankle Instability Population and Copers. *Phys. Act. Health* **2024**, *65*, 15–32. [CrossRef]
49. Gehring, D.; Faschian, K.; Lauber, B.; Lohrer, H.; Nauck, T.; Gollhofer, A. Mechanical instability destabilises the ankle joint directly in the ankle-sprain mechanism. *Br. J. Sports Med.* **2014**, *48*, 377–382. [CrossRef]
50. Caulfield, B.; Crammond, T.; O'Sullivan, A.; Reynolds, S.; Ward, T. Altered ankle-muscle activation during jump landing in participants with functional instability of the ankle joint. *J. Sport Rehabil.* **2004**, *13*, 189–200. [CrossRef]

Disclaimer/Publisher's Note: The statements, opinions and data contained in all publications are solely those of the individual author(s) and contributor(s) and not of MDPI and/or the editor(s). MDPI and/or the editor(s) disclaim responsibility for any injury to people or property resulting from any ideas, methods, instructions or products referred to in the content.

bioengineering

MDPI

Article

A Novel Approach to Predict the Location and Fatigue Life of Intervertebral Disc Degeneration

Zanni Zhang [1], Taoxi Wang [2], Huwei Bian [3], Xing Shen [2], Minjun Liang [1,*], Ee-Chon Teo [1] and Tao Jiang [3,*]

1 Faculty of Sports Science, Ningbo University, Ningbo 315211, China; zhangzanni@gmail.com (Z.Z.); eechon-teo@nbu.edu.cn (E.-C.T.)
2 State Key Laboratory of Mechanics and Control for Aerospace Structures, Nanjing University of Aeronautics and Astronautics, 29 Yudao Street, Nanjing 210016, China; wa0003xi@nuaa.edu.cn (T.W.); shenx@nuaa.edu.cn (X.S.)
3 Department of Orthopaedics, Changzhou Hospital of Traditional Chinese Medicine Affiliated to Nanjing University of Chinese Medicine, Changzhou 213000, China; doctorbhw@njucm.edu.cn
* Correspondence: liangminjun@nbu.edu.cn (M.L.); doctorjt88@163.com (T.J.)

Abstract: This study presents a novel approach for predicting the location and fatigue life of degenerative intervertebral discs (IVDs) under cyclic loading conditions, aiming to improve the understanding of disc degeneration mechanisms. Based on mechanical theories linking IVD degeneration to stress imbalance and water loss, a finite element (FE) model of the L4–L5 lumbar spine was developed, combining probability-weighted anatomical structures, inverse dynamics, and cumulative fatigue mechanics. By quantifying stress variations and calculating cumulative damage across disc regions, stress-concentration areas prone to degeneration were identified, and validation via a case study of a retired weightlifter diagnosed with intervertebral disc disease (IVDD) demonstrated that the predicted degeneration location correlated well with affected areas observed in CT scan images. These findings suggest that prolonged, abnormal stress imbalances within the disc may contribute significantly to degeneration, offering potential clinical applications in preventive assessment and targeted treatment for spine health.

Keywords: intervertebral disc disease; stress imbalance; the direction of stress attenuation; statics; fatigue mechanics

1. Introduction

Chronic low back pain (CLBP) affects up to 58% of the population [1] and is primarily caused by intervertebral disc degeneration (IVDD). Stress imbalance and water loss weaken the intervertebral disc (IVD) [2,3]. This degeneration frequently occurs at the crucial load-bearing region of the lumbar L4–L5 segment (Figure 1), studies have found that in patients with chronic low back pain, L4–L5 segments are the most common site of degenerative disc disease, and L4–L5 is located in the lumbo–sacral transition area, which carries the redistribution of upper body weight and axial pressure, and daily activities (such as sitting and bending) can create constant stress here. High-impact exercise (such as weightlifting) further increases the mechanical stress of this segment through shear and rotational forces, leading to microdamage and an accelerated degeneration of the annulus [4].

https://doi.org/10.3390/bioengineering12040423

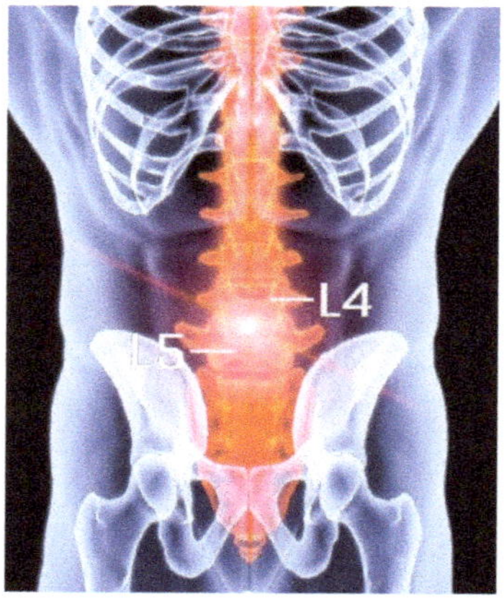

Figure 1. Location of the L4–L5 lumbar vertebrae in the human body.

The IVD consists of the nucleus pulposus (NP), annulus fibrosus (AF), and cartilage endplate [5]. The NP has a gel-like constituent that helps it to withstand compressive forces, while the AF, made of concentric collagen layers, provides tensile strength and stability [6]. These distinct material properties enable the IVD to bear loads and maintain spinal motion; however, under prolonged or excessive loading, the IVD structures undergo irreversible damage, leading to degenerative changes over time [7].

FEA has proven valuable in analysing the biomechanical behaviour of both healthy and degenerated discs [8]. An FE study simulating stress and strain within the L4–L5 segment demonstrated that degeneration significantly reduces disc stiffness, especially in the annulus fibrosus [9]. To replicate these mechanical properties, hyperplastic material models, developed based on the Mooney–Rivlin model, are commonly adopted to effectively capture the non-linear and anisotropic characteristics of the IVD [10]. Existing models predominantly focus on global stress distribution, failing to quantify how cumulative micro-damage in specific IVD regions (e.g., annulus fibrosus vs. nucleus pulposus) initiates and propagates degeneration under cyclic loading. Conventional FEA often employs simplified static or sinusoidal loads, overlooking the biomechanical complexity of high-magnitude, patient-specific dynamic activities (e.g., weightlifting) that exacerbate IVD fatigue.

Static analysis provides valuable insights into stress distribution under specific loading conditions, offering a baseline understanding of how forces are managed within the IVD during various postures and movements and the corresponding von Mises stress distributions in various regions, identifying localised stress concentration areas prone to degenerative changes [11]. Coupled with further dynamic analysis, the fatigue life prediction of localized regions, considering cumulative stress leading to progressive tissue damage, is thus determined.

Studies on lumbar spine fatigue have shown that cumulative damage from repeated compressive loads is a major factor contributing to non-traumatic vertebral fractures [11]. However, we acknowledge that compressive models are inherently limited, as intervertebral discs are subjected to complex three-dimensional stresses in vivo, including shear and torsional loads. Researchers have observed that the fatigue life of vertebral trabecular bone under axial compression can be estimated using stress normalized by volume fraction and other structural parameters [12]. These findings indicate that stress imbalances within the IVD are crucial contributors to fatigue damage accumulation, highlighting the importance of accurately predicting areas of high stress within the disc [13]. The extrapolation of fatigue mechanics frameworks, while robust in engineering contexts, demonstrates significant limitations when applied to IVDD research, most notably the absence of comprehensive spatial correlation with clinically documented patterns of localized tissue degeneration.

Accordingly, for this study, we introduced a novel approach for predicting the location and fatigue life of IVDD in the L4–L5 segment. By integrating FEA with inverse dynamics and probability-based statistical methods, we aimed to simulate stress distribution and fatigue behaviour in both normal and degenerated discs. The predictive accuracy of the methodology was evaluated through a case study involving a retired weightlifter diagnosed with IVDD following repeated 125 kg snatch loads. The model demonstrated strong agreement with clinical observations, both in identifying the degeneration location (validated against CT imaging) and quantifying fatigue life. These findings suggest that integrating biomechanical simulations with patient-specific loading profiles could enable pre-emptive identification of at-risk spinal segments, thereby informing targeted prevention strategies for IVDD.

2. Materials and Methods

2.1. Modelling of L4–L5 Segment

The L4–L5 segmental geometry was reconstructed from high-resolution CT scans of a 24-year-old male weightlifter, height 170 cm and mass 71 kg, who experienced IVD degenerative changes in the L4–L5 segment, as shown in Figure 2, due to long-term high-intensity training.

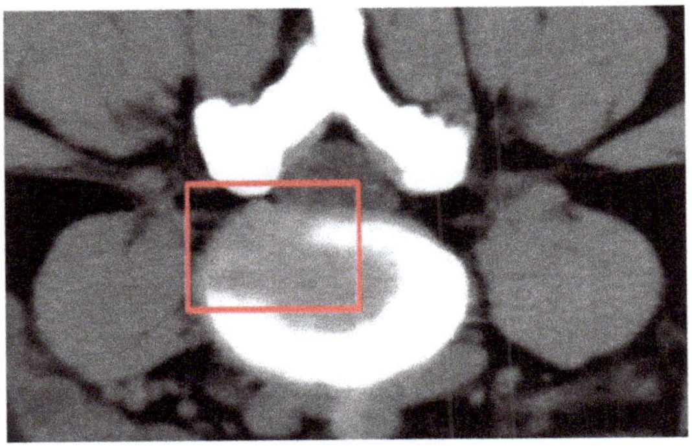

Figure 2. A CT scan image of the degenerated IVD (boxed region: degenerated disc).

A comprehensive explanation of the FE mesh generation can be found in the literature [8]. Briefly, based on 553 images of the CT scans of the lower lumbar spine, 2D images of the L4–L5 segment were cropped in three orthogonal planes to enclose the appropriate boundary for segmentation and post-processing to generate the 3D geometric solid model of L4–L5 motion segment using the contour editing tool of MIMICS (Materialize NV, Belgium). The solid model was then imported into ABAQUS 2021 to generate the FE mesh model. For the intervertebral articulating joints, the contact regions were subdivided into smaller volumes to generate two layers of elements. Different grid sizes of the various spinal components were set in the range of 2 mm to 0.6 mm. The L4–L5 FE model was generated with 74,277 nodes and 366,599 independent elements, as shown in Figure 3.

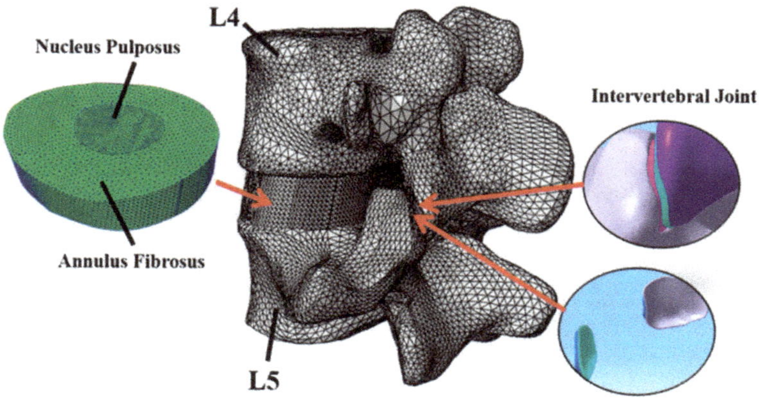

Figure 3. FE model of the L4–L5 motion segment.

2.2. Materials Properties

2.2.1. Bony Vertebrae

Based on the relationships between the apparent density (ρ) of the bony structures, the greyscale value (Hu), and elastic modulus (E), the CT grayscale values were mapped to bony material properties using empirical relationships between apparent density and elastic modulus [11,14] based on formulae:

$$\rho = 47 + 1.122\text{Hu} \quad (1)$$

$$\text{E} = 0.63 \times \rho^{1.35} \quad (2)$$

resulting in a nonuniform distribution of the material properties of the meshed bony L4–L5 segment and treated as isotropic materials with a Poisson's ratio (υ) of 0.3.

2.2.2. Soft Tissues: Annulus Fibrous and Nucleus Pulposus

The nonlinear behaviour of the nucleus pulposus and annulus fibrosus was modelled using the Mooney–Rivlin hyper-elastic material model [7], Equation (3). The Mooney–Rivlin hyper-elastic model was selected for its demonstrated capability to capture the nonlinear, anisotropic stress–strain behaviour of the annulus fibrosus under multiaxial loading conditions, while maintaining an optimal balance between predictive accuracy and parameter identifiability using experimentally derived data. The material parameters are listed in Table 1 [15–17].

$$W = c_{10}\left(\overline{I_1} - 3\right) + c_{01}\left(\overline{I_2} - 3\right) + \left(J^{el} - 1\right)^2/d \quad (3)$$

where W represents the strain potential energy while c_{01} and c_{10} represent the hyper-elastic material parameters related to the deformation of the material. I_1 and I_2 are the 1st and 2nd invariants of the deviatoric part of the right Cauchy–Green strain tensor while d and J^{el} are the material incompressibility parameter and elastic volume ratio, respectively.

Table 1. The parameter settings of hyper-elastic material in the annulus fibrosus and nucleus pulposus.

Tissues and Structures	c_{10} (MPa)	c_{01} (MPa)	d (MPa^{-1})	υ
Annulus Fibrosus	0.56	0.14	1.309	0.45
Nucleus Pulposus	0.12	0.09	17	0.49

2.2.3. Soft Tissues: Ligaments

Six primary ligaments (anterior longitudinal ligament, posterior longitudinal ligament, ligamentum flavum, interspinous ligament, supraspinous ligament, and transverse ligament) were modelled as nonlinear spring elements [7] with stiffness values as listed in Table 2 [7,18].

Table 2. The ligament stiffness settings of L4–L5 lumbar vertebrae.

Type of Ligament	Anterior Longitudinal Ligament	Posterior Longitudinal Ligament	Ligament Flava	Intertransverse Ligament	Interspinous Ligament	Supraspinal Ligament
Stiffness (N/mm)	8.74	5.83	15.38	0.19	10.85	2.39

After completing the material assignment of the L4–L5 lumbar vertebrae, the 3D model is imported into the ABAQUS for validation based on physiological loading against normal L4–L5 segments [8].

2.3. Loading and Boundary Conditions

The biomechanical evaluation of the L4–L5 segment under cyclic loading encompassed a combined static–dynamic analytical framework to comprehensively characterize both baseline stress distributions and cumulative fatigue damage progression. This approach enabled the identification of stress-concentration zones and the prediction of fatigue life under physiologically relevant loading conditions.

2.3.1. Static Stress Analysis

Static analysis was conducted to evaluate stress distribution within the segment under compressive and moment loading. A compressive load of 500 N and a moment of 10 Nm were applied to the superior surface of L4 (These selected loading conditions—500 N axial compression and 10 Nm flexion–extension moment—were based on biomechanical studies quantifying lumbar spinal loads during activities of daily living and weightlifting. The 500 N compressive force approximates the upper body weight transmitted through the lumbar spine during upright standing, while the 10 Nm moment reflects moderate lumbar flexion observed in lifting postures, as validated by in vivo intradiscal pressure measurements), while the inferior surface of L5 was fully constrained [7]. The IVD was divided into nine equal-volume regions (Figure 4), and the computed von Mises stress in each region was extracted. Quantitative difference (QD1) and coefficient of variation (CV) methods were applied to identify stress levels [19,20]. The spatial heterogeneity of stress distribution within the IVD under cyclic loading was systematically evaluated using two complementary metrics: the QD1 and the CV. The QD1 metric quantifies relative stress disparities between adjacent IVD regions by comparing localized von Mises stresses to a reference region. This approach highlights regions with abnormally elevated mechanical gradients, which are strongly associated with microdamage initiation and clinical degeneration patterns. For instance, a high QD1 value in the annulus fibrosus indicates disproportionate stress concentrations relative to surrounding tissues, identifying potential fatigue hotspots. In parallel, the CV metric assesses intra-regional stress variability by calculating the ratio of the standard deviation to the mean stress within a partitioned region. Elevated CV values signify pronounced stress fluctuations, often linked to microstructural inhomogeneity or transient loading peaks. Such variability exacerbates cumulative fatigue damage, particularly in mechanically vulnerable regions like the posterolateral annulus.

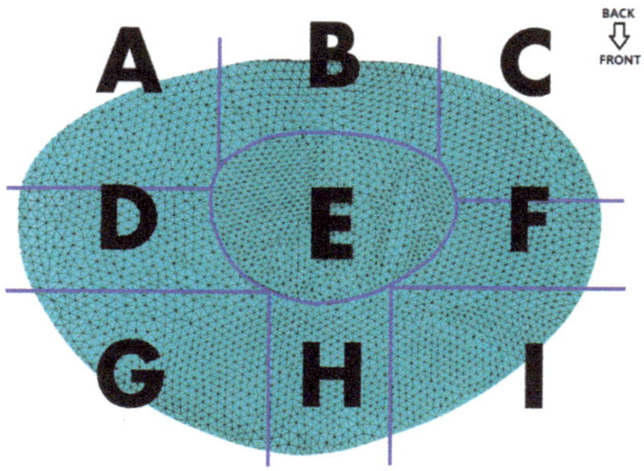

Figure 4. The illustration of partition in the IVD. The IVD was divided into nine regions (A–I), representing posterior annulus fibrosus (A–C), center regions (D–F), and anterior (G–I) anterior annulus fibrosus from left to right.

In the FE analysis, a reference point (RP) was established 5 mm superior to the L4 vertebral body to serve as a centralized load application node. This RP was kinematically coupled to the entire superior surface of L4 using a distributing coupling constraint, which ensures uniform force transmission by linearly mapping displacements and rotations at the RP to all nodes on the L4 superior surface. The coupling constraint was implemented through a multi-point constraint (MPC) formulation, where the rigid body motion of the RP governs the deformation of the L4 surface while allowing local elastic deformations within physiological limits. This approach mimics the load transfer mechanism observed in vivo, where forces are distributed across the vertebral endplate via the intervertebral disc's composite structure (Figure 5).

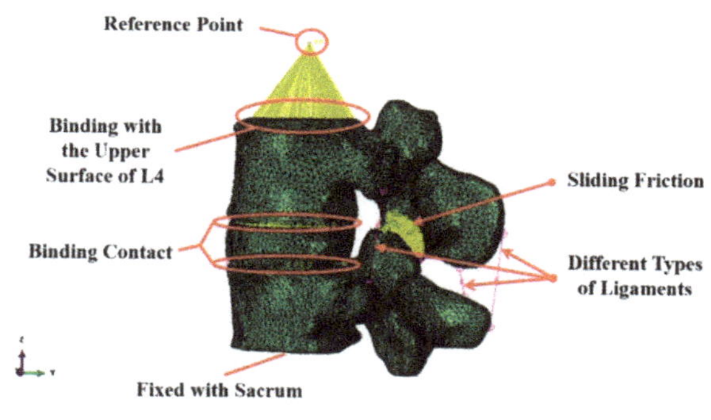

Figure 5. FEM model of L4–L5 lumbar vertebrae.

2.3.2. Dynamic Analysis

Dynamic fatigue analysis was performed using transient dynamics simulations in ANSYS R16 (ANSYS, Inc., Canonsburg, PA, USA) [5,21]. Loading conditions were derived from inverse dynamics simulations of a 125 kg snatch performed by the weightlifter, modelled in AnyBody [22,23]. The inverse dynamics framework employed in this study is based on the Newton–Euler equations:

$$\tau = M(q)\ddot{q} + C(q,\dot{q}) + G(q) + J\left(q\right)^T F_{ext} \quad (4)$$

where the joint torque vector (τ, in N·m) is calculated as the sum of inertial ($M(q)\ddot{q}$), coriolis and centrifugal force vector $C(q,\dot{q})$, gravitational force vector $G(q)$, and external force $J(q)^T F_{ext}$ components, with $M(q)$ representing the configuration-dependent mass matrix (kg·m^2), $\ddot{q}$ denoting joint angular accelerations (rad·s^2) derived from motion capture data, $C(q,\dot{q})$ accounting for velocity-dependent coupling effects, $G(q)$ incorporating segmental gravitational forces, and $J(q)^T F_{ext}$ translating the external 125 kg barbell load to joint space through the system's Jacobian matrix.

The fatigue life of the most stressed regions was calculated using Miner's cumulative damage law:

$$N_f = \frac{C}{\sigma_a^m} \tag{5}$$

where σ_a^m is the stress amplitude, N_f is the number of cycles to failure, and C and m are material constants. MATLAB's Rainflow counting algorithm processed stress-time data, generating stress-cycle distributions for fatigue life prediction [11,14,18].

2.4. Fatigue Analysis

2.4.1. Fatigue Life Prediction

Fatigue analysis was performed using Miner's cumulative damage rule to calculate the cumulative damage value for each region:

$$D = \sum_{i=1}^{k} \frac{n_i}{N_i} \tag{6}$$

For each partitioned disc section, stress-time histories from dynamic simulations were decomposed into discrete stress cycles using Rainflow counting. The number of cycles at each stress level (n_i) was normalized by the corresponding fatigue life (N_i)—defined as the cycles to failure at that stress amplitude—yielding incremental damage values. Cumulative damage (D) was calculated as the sum of these increments across all stress levels [7]. Failure is assumed when $D \geq 1$ [12,24].

2.4.2. Stress Cycle Extraction

With the stress–time data under dynamic loading obtained from transient dynamic analysis, MATLAB's (MATLAB R2022a MathWorks, Inc., Natick, MA, USA) Rainflow algorithm was used to extract stress cycles, identifying stress amplitudes and cycle distributions at key nodes [14]. The algorithm decomposes complex loading histories into discrete stress cycles by detecting hysteresis loops, enabling the quantification of both peak-to-peak amplitudes and mean stress levels critical for fatigue life prediction.

3. Results

3.1. Fatigue Hotspots

Stress-cycle distributions revealed posterior regions of the IVD, particularly sections F, H, and I, experienced the greatest stress amplitudes, demonstrating significant cumulative damage trends (Table 3, Figures 6 and 7), the use of von Mises stress as a predictor of mechanical vulnerability in the annulus fibrosus aligns with prior FE studies demonstrating its correlation with microstructural failure under compressive loading [17]. Segment G is located posterolateral to the annulus fibrosus, near the pedicle–annulus interface, and has low cumulative damage relative to the H/I cross-section.

Table 3. The QD1 results of von Mises stress in different regions of the IVD.

Location	von Mises (MPa)		CV	QD1	
	Mean $\bar{x}$	Std.		Mean $\bar{x}$	CV
Section A	0.593	0.191	0.321	1.633 ++	1.219 +
Section B	0.624	0.244	0.391	1.741 ++	1.625 ++
Section C	0.741	0.432	0.583	2.095 ++	2.456 ++
Section D	0.748	0.293	0.392	2.117 ++	1.631 ++
Section E	0.270	0.048	0.179	0.000	0.000
Section F	0.977	0.469	0.480	2.671 ++	2.053 ++
Section G	0.989	0.416	0.420	2.697 ++	1.776 ++
Section H	1.157	0.652	0.564	3.022 ++	2.388 ++
Section I	1.157	1.108	0.957	3.022 ++	3.487 ++

+: QD1 ≥ 1.5 (moderate stress deviation from reference region). ++: QD1 ≥ 2.0 (severe stress deviation from reference region).

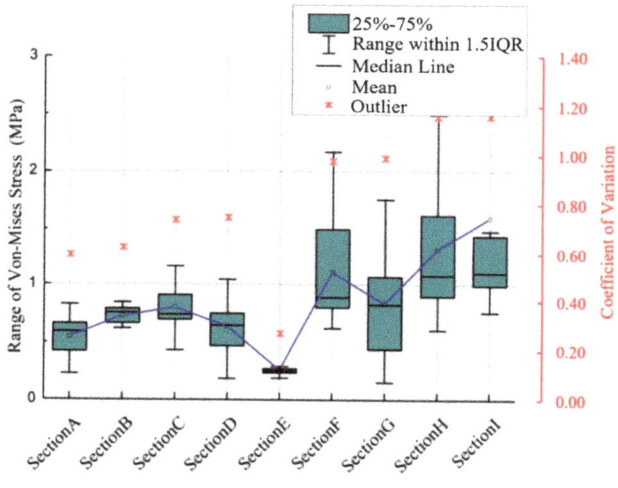

Figure 6. The range and CV of the von Mises stress in different sections of the disc in QD1.

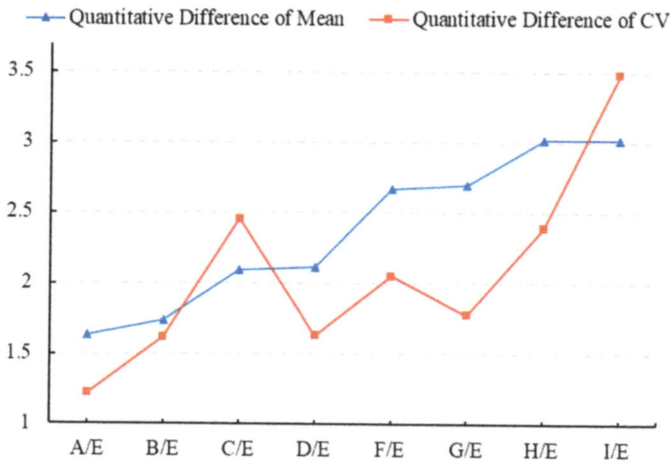

Figure 7. The QD1 results of mean and CV between section E and other sections in the disc.

Figure 8 illustrates the spatial heterogeneity of von Mises stress within the IVD, highlighting significant differences between the annulus fibrosus and nucleus pulposus, as well as adjacent subregions. The annulus fibrosus showed significantly higher stress amplitudes, with greater stress in the posterolateral region (H and I) than in the anterolateral region.

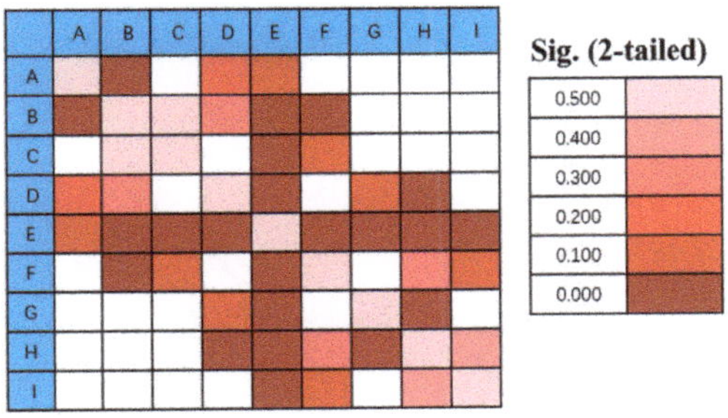

Figure 8. The pairwise differences in von Mises stress between adjacent regions of the IVD, labelled A through I. The colour intensity corresponds to the magnitude of the stress difference, with darker red indicating smaller differences (closer to 0.0) and lighter pink indicating larger differences (up to 0.5), as shown in the legend on the right.

According to the statistical results and based on the mechanics concept, the computed mean value of von Mises stress in the E area (nucleus pulposus) is the smallest. The average von Mises stresses in regions F, H, and I of the annulus fibrosus are generally higher than those in regions C and G, which in turn are greater than those in regions A, B, and D. This pattern indicates a gradient of stress attenuation across the annulus fibrosus, decreasing from the right anterior to the left posterior direction, as illustrated in Figure 9. The figure depicts the spatial gradient of mechanical stress dissipation within the intervertebral disc, characterized by a progressive reduction in von Mises stress from high-load regions (e.g., the posterolateral annulus) toward low-load regions (e.g., the nucleus pulposus). This directional pattern reflects the disc's load-transfer mechanism, where the anisotropic structure of the annulus fibrosus causes stress to dissipate along specific anatomical pathways. In our study, stress attenuation followed a consistent trajectory from the right anterior to left posterior annulus, correlating with clinical observations of degeneration patterns and demonstrating how disc anatomy governs mechanical vulnerability, and "A" is the section being squeezed.

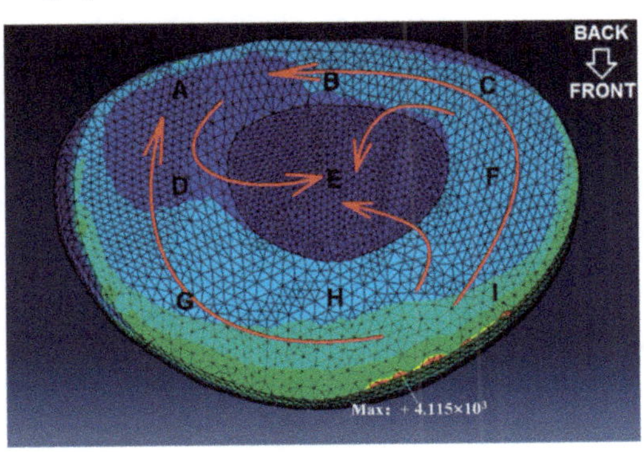

Figure 9. The direction of stress attenuation in the IVD. The colour gradient follows a standardized convention where red hues denote high values (elevated von Mises stress), blue hues represent low values, and intermediate colours (green/yellow) indicate mid-range quantities.

3.2. Fatigue Life of the Section Being Squeezed

The Rainflow counting algorithm results (Figure 10) showed that anterior regions (F, H, and I) experienced a disproportionately high number of stress cycles with amplitudes exceeding 1.0 MPa. These anterior annulus fibrosi endured substantial tensile stresses and shear forces during repetitive dynamic loading. Fatigue life predictions indicate section I had the shortest lifespan, aligning with clinical observations of degeneration commonly originating in anterior-lateral disc regions.

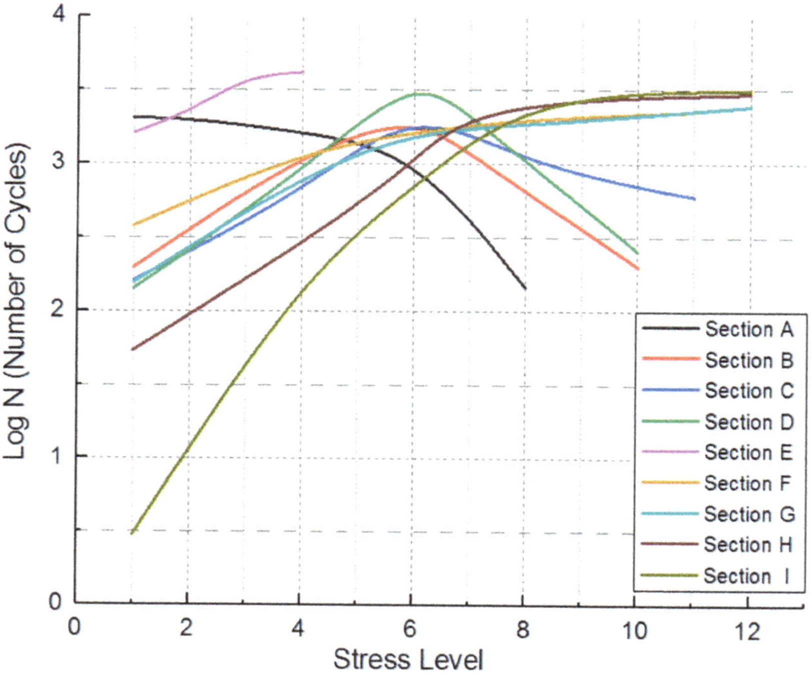

Figure 10. Distribution of stress levels versus the number of cycles in different sections.

Conversely, section A (posterior-left) demonstrates a significantly longer fatigue life, which can be attributed to its thicker fibrous structure and reduced exposure to tensile stresses. This result highlights the structural advantage of the anterior annulus fibrosus in resisting mechanical fatigue. The findings underscore the importance of stress amplitude and variability in predicting degeneration-prone regions, providing a quantitative framework for early diagnosis and preventive care in spinal health.

The predicted results align with the CT imaging data of the retired weightlifter (Figure 2), validating the accuracy of the proposed method. The combined analysis of regional stress distributions (Figure 8) and fatigue life predictions (Figure 10) demonstrated that posterior annulus fibrosus regions (Sections F, H, I) exhibited sustained stress amplitudes exceeding 1.0 MPa—a threshold associated with collagen microfailure—resulting in a 3.2-fold reduction in predicted fatigue life compared to nucleus pulposus regions (Section E, 0.28 $\pm$ 0.05 MPa). These findings quantitatively establish stress imbalance as a dominant mechanical driver of localized IVD degeneration.

Table 4 stratifies stress amplitude levels across IVD regions under cyclic loading, correlating them with predicted fatigue life. This classification enables a quantitative assessment of mechanical risk zones within the disc, providing a foundation for understanding how local stress magnitudes influence tissue degeneration and potential failure. By assigning discrete stress levels, the table facilitates comparative analysis across regions and supports fatigue-based modelling of disc mechanics under repetitive loading conditions.

Table 4. The stress amplitude level division of IVD used in Figure 10.

Stress Amplitude (MPa)	Level	Stress Amplitude (MPa)	Level
1.1~1.2	12	0.5~0.6	6
1.0~1.1	11	0.4~0.5	5
0.9~1.0	10	0.3~0.4	4
0.8~0.9	9	0.2~0.3	3
0.7~0.8	8	0.1~0.2	2
0.6~0.7	7	0.0~0.1	1

4. Discussion

This study bridges key concepts from materials science and bone biomechanics to advance IVD degeneration prediction. Like the fatigue damage modelling in structural alloys under complex loading [25], our framework quantified cumulative mechanical damage in biological tissues by adapting Miner's law to multiaxial stress states, demonstrating that annulus fibrosus collagen exhibits fatigue failure patterns analogous to engineered materials. The "theramechanics" paradigm [26], which advocates mechanically informed therapies, directly supports our proposal to translate fatigue life predictions into personalized activity prescriptions—shifting from passive imaging-based diagnosis to proactive biomechanical optimization. Furthermore, the coupling of damage evolution with diffusive bio-mechanical signals in bone remodelling [27] parallels our findings on stress-driven nutrient transport alterations in degenerated discs, suggesting a universal mechanism where mechanical insults disrupt tissue homeostasis.

Prior studies of IVDD have predominantly focused on static stress distributions or simplified cyclic loading scenarios, neglecting patient-specific dynamic loading patterns (e.g., weightlifting manoeuvres) and their cumulative fatigue effects. This oversight limits the ability to predict where and when degeneration initiates under real-world mechanical conditions. The clinical urgency to address IVDD-related disability, coupled with biomechanical evidence linking stress imbalance to microstructural failure, necessitated a framework integrating dynamic loading profiles, anatomic variability, and fatigue mechanics. This approach directly addresses the unmet need for spatially resolved, activity-specific degeneration risk prediction. Therefore, this study introduced a novel approach for predicting IVD degeneration using a combination of FEA, inverse dynamics, and fatigue mechanics. Results demonstrated that stress concentrations and fatigue life predictions align with clinical observations, supporting the hypothesis that long-term stress imbalance contributes to IVD degeneration. The verification study quantified regional variations in mean von Mises stress and fatigue life within the IVD. Under 125 kg snatch loading conditions, section A demonstrated maximal fatigue resistance, whereas section I displayed minimal durability. The IVD experienced the highest von Mises stress on the anterior lateral right (section I) which was gradually attenuated towards the posterior lateral left (section A).

The model's predictive accuracy was quantitatively assessed through spatial correlation analysis between predicted high-stress regions and CT-identified degeneration sites in the case study subject. A semi-automated segmentation protocol (ITK-SNAP 4.2.0 software) was employed to map degeneration loci from clinical CT scans to the FE model's coordinate system, enabling a direct comparison of spatial overlap. The posterolateral annulus regions (Sections H, I) demonstrated concordance between predicted stress concentrations and actual degeneration areas.

Since the annulus fibrosi are thicker on the anterior lateral regions than on the posterior lateral regions, and with the exact location of the nucleus pulposus slightly behind the centre of the IVD, the posterior and posterolateral annulus become weaker mechanically. The anterior part of the IVD mainly bears the compressive stress, while the middle and posterior parts mainly bear tensile stress, and shear force is generated in the IVD. The long-term stress imbalance during weightlifting may have caused the pressure difference in the disc and led to the disc being squeezed from high pressure to low pressure. Hence, the section being squeezed in IVD occurs in the left posterior part of the disc (section A), thus possibly accelerating the degeneration of the IVD under long-term stretching, leading to IVDD. By comparing the CT images (as shown in Figure 2), the model's predicted degeneration zones showed excellent spatial correspondence with radiologically confirmed IVDD regions, corroborating the biomechanical simulation's validity. Meanwhile, based on the mechanical theory of IVDD pathogenesis, the data suggest that the long-term stress imbalance inside the IVD may be the potential reason for IVDD; the approach on the basis of such theory for location prediction is feasible.

The biomechanical model proposed in this study provides clinicians with a decision support tool that goes beyond traditional imaging assessments. By predicting the fatigue life and stress distribution of intervertebral discs under functional load, this method can (1) identify high-risk degeneration areas before morphological changes occur and achieve early intervention; (2) quantify the impact of different activities on the degenerative process and provide patients with personalized exercise prescriptions; and (3) complement traditional examination methods such as MRI, upgrading static imaging evaluation to dynamic functional analysis. This mechanics-based assessment framework enables clinicians to move from passive diagnosis to proactive prevention and is particularly suitable for spinal health management in athletes and heavy manual workers.

This study is limited by its reliance on a single case study. Variability in patient-specific factors, such as age, body mass, and activity level, may influence the generalizability of the results. Furthermore, the current model applies only compressive loading, which does not fully replicate the complex in vivo loading conditions experienced by the intervertebral disc. In reality, the IVD is subjected to multi-axial forces including shear and torsional loads, which may significantly affect stress distribution and fatigue behaviour. Future work should therefore incorporate a more comprehensive loading regime and include a larger cohort to validate the model across different patient demographics and physical activities.

5. Conclusions

This study presents a predictive model to identify high-stress areas in the IVD, offering insights for fatigue life predictions, injury prevention, and spinal health monitoring. This proposed approach, based on mechanical theories of stress imbalance, integrating FEA and mathematical statistics to predict IVDD positions, was validated through practical application.

The findings emphasize the critical role of stress imbalance in IVD degeneration, offering a potential tool for early diagnosis and injury prevention. By quantifying stress distribution and fatigue life, the model can support personalized treatment planning and preventive strategies in spine health management. Through this novel approach, the location of the IVD degeneration and herniation can be predicted and quantified, and medical guidance for injury prevention can be given in advance, to reduce the risk of IVDD. Future work may focus on validating this model across varied physical activities and integrating patient-specific parameters to enhance clinical applicability in preventative care for IVDD.

Author Contributions: Conceptualization, M.L., E.-C.T. and T.J.; methodology, T.W. and X.S.; software, Z.Z. and X.S.; validation, H.B., T.J. and Z.Z.; formal analysis, Z.Z.; investigation, Z.Z. and T.W.; resources, M.L. and E.-C.T.; data curation, Z.Z. and H.B.; writing—original draft preparation, Z.Z.; writing—review and editing, M.L., E.-C.T. and T.J.; visualization, X.S.; supervision, M.L., E.-C.T. and T.J.; project administration, M.L.; funding acquisition, T.J. All authors have read and agreed to the published version of the manuscript.

Funding: This project was supported by the Key Project of the Science and Technology Development Plan for Traditional Chinese Medicine in Jiangsu Province (Grant No. ZD202325).

Institutional Review Board Statement: All research procedures were approved by the Medical Ethics Committee of Ningbo University (approval number: TY2024050) and comply with the provisions of the Declaration of Helsinki.

Informed Consent Statement: Participants provided informed written consent prior to participating in this study.

Data Availability Statement: The original contributions presented in this study are included in the article. Further inquiries can be directed at the corresponding authors.

Conflicts of Interest: The authors declare no conflicts of interest, and the manuscript is approved by all authors for publication.

References

1. Hoy, D.; Brooks, P.; Blyth, F.; Buchbinder, R. The Epidemiology of low back pain. *Best Pract. Res. Clin. Rheumatol.* **2010**, *24*, 769–781. [CrossRef] [PubMed]
2. Tao, C.; Lin, S.; Shi, Y.; Gong, W.; Chen, M.; Li, J.; Zhang, P.; Yao, Q.; Qian, D.; Ling, Z.; et al. Inactivation of Tnf-α/Tnfr signaling attenuates progression of intervertebral disc degeneration in mice. *JOR Spine* **2024**, *7*, e7000. [CrossRef] [PubMed]
3. Park, J.S.; Goh, T.S.; Lee, J.S.; Lee, C. Analyzing isolated degeneration of lumbar facet joints: Implications for degenerative instability and lumbar biomechanics using finite element analysis. *Front. Bioeng. Biotechnol.* **2024**, *12*, 1294658. [CrossRef]
4. Varga, M.; Štulík, J.; Pivarči, F.; Geri, G.; Malík, J.; Lang, O.; Srikandarajah, N.; Kaiser, R. Correlation of MRI-Evaluated Degenerative Disc Disease with Positivity on Single-Photon Emission Computed Tomography Imaging in Patients with Chronic Low Back Pain. *World Neurosurg.* **2024**, *190*, e504–e512. [CrossRef] [PubMed]
5. Benneker, L.M.; Heini, P.F.; Anderson, S.E.; Alini, M.; Ito, K. Correlation of radiographic and MRI parameters to morphological and biochemical assessment of intervertebral disc degeneration. *Eur. Spine J. Off. Publ. Eur. Spine Soc. Eur. Spinal Deform. Soc. Eur. Sect. Cerv. Spine Res. Soc.* **2005**, *14*, 27–35. [CrossRef]
6. Lee, C.K.; Kim, Y.E.; Lee, C.S.; Hong, Y.M.; Jung, J.M.; Goel, V.K. Impact response of the intervertebral disc in a finite-element model. *Spine* **2000**, *25*, 2431–2439. [CrossRef]
7. Schmidt, H.; Kettler, A.; Heuer, F.; Simon, U.; Claes, L.; Wilke, H.J. Intradiscal pressure, shear strain, and fiber strain in the intervertebral disc under combined loading. *Spine* **2007**, *32*, 748–755. [CrossRef]
8. Geng, Y.; Wang, T.; Xu, C.; Bian, H.; Shen, X.; Teo, E.-C.; Jiang, T. Kinematic Responses and Segmental Stiffness of Normal and Degenerative L4/L5 Motion Segments: Finite Element Study. *J. Mech. Med. Biol.* **2024**, *24*, 2450018. [CrossRef]
9. Rohlmannt, A.; Claes, L.E.; Bergmannt, G.; Graichen, F.; Neef, P.; Wilke, H.J. Comparison of intradiscal pressures and spinal fixator loads for different body positions and exercises. *Ergonomics* **2001**, *44*, 781–794. [CrossRef]
10. Wilson, S.P.; Taylor, D. Statistical analysis and reliability prediction with short fatigue crack data. *Fatigue Fract. Eng. Mater. Struct.* **1999**, *22*, 67–76. [CrossRef]
11. Haddock, S.M.; Yeh, O.C.; Mummaneni, P.V.; Rosenberg, W.S.; Keaveny, T.M. Similarity in the fatigue behavior of trabecular bone across site and species. *J. Biomech.* **2004**, *37*, 181–187. [CrossRef]
12. Rapillard, L.; Charlebois, M.; Zysset, P.K. Compressive fatigue behavior of human vertebral trabecular bone. *J. Biomech.* **2006**, *39*, 2133–2139. [CrossRef] [PubMed]
13. Chen, S.H.; Tai, C.L.; Lin, C.Y.; Hsieh, P.H.; Chen, W.P. Biomechanical comparison of a new stand-alone anterior lumbar interbody fusion cage with established fixation techniques—A three-dimensional finite element analysis. *BMC Musculoskelet. Disord.* **2008**, *9*, 88. [CrossRef]
14. Liu, Y.K.; Njus, G.; Buckwalter, J.; Wakano, K. Fatigue response of lumbar intervertebral joints under axial cyclic loading. *Spine* **1983**, *8*, 857–865. [CrossRef] [PubMed]
15. Rho, J.Y.; Kuhn-Spearing, L.; Zioupos, P. Mechanical properties and the hierarchical structure of bone. *Med. Eng. Phys.* **1998**, *20*, 92–102. [CrossRef]

16. Smit, T.H.; Odgaard, A.; Schneider, E. Structure and function of vertebral trabecular bone. *Spine* **1997**, *22*, 2823–2833. [CrossRef] [PubMed]
17. Goel, V.K.; Monroe, B.T.; Gilbertson, L.G.; Brinckmann, P. Interlaminar shear stresses and laminae separation in a disc. Finite element analysis of the L3-L4 motion segment subjected to axial compressive loads. *Spine* **1995**, *20*, 689–698. [CrossRef]
18. Chen, C.S.; Cheng, C.K.; Liu, C.L.; Lo, W.H. Stress analysis of the disc adjacent to interbody fusion in lumbar spine. *Med. Eng. Phys.* **2001**, *23*, 483–491. [CrossRef]
19. Lin, C.C.; Yang, C.P.; Li, C.I.; Liu, C.S.; Chen, C.C.; Lin, W.Y.; Hwang, K.L.; Yang, S.Y.; Li, T.C. Visit-to-visit variability of fasting plasma glucose as predictor of ischemic stroke: Competing risk analysis in a national cohort of Taiwan Diabetes Study. *BMC Med.* **2014**, *12*, 165. [CrossRef]
20. Krinsley, J.S.; Egi, M.; Kiss, A.; Devendra, A.N.; Schuetz, P.; Maurer, P.M.; Schultz, M.J.; van Hooijdonk, R.T.; Kiyoshi, M.; Mackenzie, I.M.; et al. Diabetic status and the relation of the three domains of glycemic control to mortality in critically ill patients: An international multicenter cohort study. *Crit. Care* **2013**, *17*, R37. [CrossRef]
21. Boos, N.; Weissbach, S.; Rohrbach, H.; Weiler, C.; Spratt, K.F.; Nerlich, A.G. Classification of age-related changes in lumbar intervertebral discs: 2002 Volvo Award in basic science. *Spine* **2002**, *27*, 2631–2644. [CrossRef]
22. Damsgaard, M.; Rasmussen, J.; Christensen, S.T.; Surma, E.; de Zee, M. Analysis of musculoskeletal systems in the AnyBody Modeling System. *Simul. Model. Pract. Theory* **2006**, *14*, 1100–1111. [CrossRef]
23. Goel, V.K.; Kong, W.; Han, J.S.; Weinstein, J.N.; Gilbertson, L.G. A combined finite element and optimization investigation of lumbar spine mechanics with and without muscles. *Spine* **1993**, *18*, 1531–1541. [CrossRef] [PubMed]
24. Brinckmann, P.; Biggemann, M.; Hilweg, D. Fatigue fracture of human lumbar vertebrae. *Clin. Biomech.* **1988**, *3* (Suppl. 1), i-S23. [CrossRef] [PubMed]
25. Dell'Isola, F.; Volkov, I.A.; Igumnov, L.A.; Eugster, S.R.; Litvinchuk, S.Y.; Kazakov, D.A.; Gorohov, V.A.; Abali, B.E. Estimating fatigue related damage in alloys under block-type non-symmetrical low-cycle loading. *New Achiev. Contin. Mech. Thermodyn. Tribut. Wolfgang H. Müller* **2019**, *108*, 81–92.
26. Allena, R.; Rémond, Y. Theramechanics: How acting on mechanics will help conceive new medical treatments. *Math. Mech. Complex Syst.* **2023**, *11*, 541–566. [CrossRef]
27. Addessi, D.; D'Annibale, F.; Placidi, L.; Giorgio, I. A bone remodeling approach encoding the effect of damage and a diffusive bio-mechanical stimulus. *Contin. Mech. Thermodyn.* **2024**, *36*, 993–1012. [CrossRef]

Disclaimer/Publisher's Note: The statements, opinions and data contained in all publications are solely those of the individual author(s) and contributor(s) and not of MDPI and/or the editor(s). MDPI and/or the editor(s) disclaim responsibility for any injury to people or property resulting from any ideas, methods, instructions or products referred to in the content.

bioengineering

MDPI

Article

Ground Reaction Forces and Impact Loading Among Runners with Different Acuity of Tibial Stress Injuries: Advanced Waveform Analysis for Running Mechanics

Ryan M. Nixon [1,*,†], Sharareh Sharififar [1,†], Matthew Martenson [1,†], Lydia Pezzullo [1], Kevin R. Vincent [2,†] and Heather K. Vincent [1,†]

1 Exercise Medicine and Functional Fitness Laboratory, Department of Physical Medicine and Rehabilitation, University of Florida, Gainesville, FL 32611, USA; sharareh75@ufl.edu (S.S.); matthew.martenson@medicine.ufl.edu (M.M.); lpezzullo@ufl.edu (L.P.); heatherketelaar@gmail.com (H.K.V.)

2 The Orthopaedics Institute, Alachua, FL 32615, USA; krvincent1@gmail.com

* Correspondence: ozswinem@ufl.edu; Tel.: +1-352-256-5923

† These authors contributed equally to this work.

Abstract

Conventional ground reaction force (GRF) and load rate (LR) analyses may overlook temporal and waveform characteristics that reflect injury status and acuity. This study used an alternative GRF processing methodology to characterize GRF waveforms among runners with symptomatic medial tibial stress fractures (MTSS) and those recovering from tibial stress fractures (TSF; both unilateral [UL] and bilateral [BL]). This cross-sectional analysis of runners ($n = 66$) included four groups: symptomatic MTSS, recovering from UL or BL TSF, or uninjured case-matched controls. Participants ran at self-selected speed on an instrumented treadmill. Kinematics were collected with a 3D optical motion analysis system. Double-Gaussian models described the biphasic loading pattern of running gait (initial impact, active phases). Gaussian parameters described relative differences in the GRF waveform by injury condition. LR was calculated using the central difference numerical derivative of the raw normalized net force data. During the impact phase (0–20% of stance), controls and BL TSF produced higher GRF amplitudes than UL TSF and MTSS ($p < 0.05$). BL TSF and controls had greater maximal positive LR and minimum LR than UL TSF and MTSS. Peak medial GRF was 18–43% higher in the BL TSF group than in MTSS and UL TSF ($p < 0.05$). Correlations existed between tibial pain severity and early stance net GRF ($r = 0.512$; $p = 0.016$) and between pain severity and the duration since diagnosis for LR values during the impact phase (r values = 0.389–0.522; all $p < 0.05$). Collectively, these data suggest that this waveform modeling approach can differentiate injury status and pain acuity in runners. Early stance GRF and LR may offer novel insight into the management of running-related injuries.

Keywords: running; tibial stress fracture; medial tibial stress syndrome; ground reaction force; load rates

1. Introduction

Medial tibial stress syndrome (MTSS) and tibial stress fractures (TSF) are among the most prevalent running-related injuries, commonly resulting from insufficient adaptation to mechanical loading [1]. These injuries represent a continuum of overuse-related tibial bone stress injuries, starting with painful MTSS (periostitis) and possibly transitioning to stress fractures in either unilateral or bilateral sites [2]. Epidemiological studies indicate that between 13% and 20% of runners experience MTSS during their training, and TSF accounts for up to 49% of all stress fractures in athletes [1,3,4]. Bilateral cases are uncommon, with bilateral TSF reported in 16% of athletes with stress fractures [5]. Injuries have the potential to disrupt training, hinder running performance, and impede recovery, thereby necessitating an enhanced understanding of the biomechanical forces that contribute to their onset and progression [6,7].

Altered ground reaction force (GRF) and load rate (LR) characteristics during loading and unloading may provide valuable insight into injury-specific responses and recovery dynamics [8].The association of loading patterns with these injuries is not consistent. Systematic reviews in this area suggest higher LRs are associated with stress fracture onset, whereas GRFs are not [9]. Other reviews do not support an association between LR and fracture [10]. These discrepant findings may be partly due to the methodology of prior studies and the collection of loading patterns at different phases of bone overuse injury. Pain, the fear of loading a recently injured limb, the duration of injury, and severity can all influence GRF responses, and these factors are not typically presented. Moreover, conventional GRF analyses focus on discrete metrics, such as peak impact forces and LR magnitudes during specific regions of the gait cycle [6,11]. While these metrics are informative, they are inadequate to capture critical temporal and waveform characteristics essential in understanding injury status, particularly in the case of MTSS and TSF. MTSS, resulting from fatigue loading and the tissue's inability to adapt or recover from repetitive stresses, may produce altered GRF waveforms with elevated impact forces and irregular loading/unloading patterns compared to healthy runners [12] potentially due to pain symptoms. Conversely, individuals recovering from TSF, especially during the early recovery stages when pain is absent, may demonstrate a conservative GRF pattern, with damped impact forces to protect the healing bone and minimize fracture risk [13]. As recovery progresses, there may be gradual normalization of GRF patterns. However, mediolateral force asymmetry could persist due to compensatory strategies or weakness in the affected limb [14]. Focusing on injury acuity among runners along the bone injury spectrum from acute pain in MTSS to the recovery stages of TSF is a valuable and understudied area. Findings may uncover critical differences in loading patterns that might influence injury onset, guide therapy design and progression, or enhance recovery.

Alternative preprocessing methods that preserve key biomechanical phenomena of the waveform by preferentially using averaging and normalization over lowpass filtering enable a more nuanced examination of the mechanical loads during a typical running gait cycle. While conventional methods often use extensive filtering and data smoothing to process running GRF signals, our approach is intended to maintain as much of the real GRF signal as possible to capture unique signals related to the acuity of bone injury. Therefore, this research aimed to characterize multidimensional GRF features, including net force, medial and lateral GRFs, full-stance LR waveforms, and force redistribution strategies across runners with symptomatic MTSS, those recovering from TSF (both unilateral [UL] and bilateral [BL]), and healthy non-injured control runners. This study applied a double-Gaussian model to analyze GRF waveforms, capturing the biphasic loading pattern inherent in running [15]. It was hypothesized that runners with symptomatic, acute MTSS would demonstrate elevated LR, increased mediolateral GRF asymmetries, and altered waveform patterns compared to healthy controls and individuals in post-recovery stages of TSF. These findings may help clarify the role of GRF and LR in running-related bone injury.

2. Materials and Methods

2.1. Study Design

This was a cross-sectional analysis of the biomechanical differences among endurance runners recovered from UL TSF or BL TSF, or with current symptomatic MTSS. A group of matched non-injured runners was selected based on sex, BMI, and years of experience. This study and its procedures followed the Declaration of Helsinki's Protection of Human Subjects guidelines and were approved by the University of Florida Institutional Review Board (IRB # 202500639).

2.2. Setting

Testing was performed in the Exercise Medicine and Functional Fitness laboratory, which is located in a quaternary health care facility.

2.3. Participants

Runners provided written informed consent for participation in our research databank (IRB # 202101632). A sample of recreational and competitive runners with a physician's diagnoses of UL TSF, BL TSF, and current symptomatic MTSS was collected from 1 January 2016, to 10 April 2025, identifying 33 runners with these specific diagnoses exclusively. A group of non-injured recreational and competitive runners was selected and matched by sex, BMI, and years of experience. Patients were excluded if: (1) they presented with other acute or chronic injury diagnoses (e.g., low back pain, patellofemoral pain, other stress fractures), (2) the participant at the time of testing was not able to run consistently for the treadmill test (unable to run for the whole testing time). A total of 66 runners were used for this analysis.

2.4. Data Collection and Measurements

Data were acquired from a comprehensive health history intake and biomechanical running gait analyses [16,17].

Health History Intake. Medical, health, and training history were collected from a comprehensive intake form based on our published recommendations for runner assessment [18]. Data collection included characteristics and medical history, any pain symptoms and pain severity (presence of pain in major joints or muscle groups, severity of which is captured during history review using the 11-point Numerical Pain Rating scale [NRSpain]), training history (volume, type, cross-training activities, strengthening exercise), shoe wear and orthotics if applicable (weight, heel-to-toe drop, heel height), and whether or not they were currently training for competition (yes, no). Each runner was permitted to use their habitual preferred shoes for the testing to minimize any acute effects of changing footwear on the kinematic and kinetic data. The months since the diagnosis of tibial injury were self-reported. For the analysis, participants were asked to rate pain severity during running using the NRS_{pain} where 0 = no pain and 10 = the most excruciating pain experienced.

Procedures and Instrumentation. Force plate data were collected from an instrumented treadmill (AMTI, Watertown, MA; USA) at 1200 Hz [18,19]. Kinematic data were collected with a 7-camera, three-dimensional motion capture system at 120 Hz (Motion Analysis, Rohnert Park, CA, USA). A total of 33 retroreflective markers were applied to anatomical landmarks from the shoulder to the foot (bilateral calcanei, the base of the hallux, the base of 5th metatarsal, medial and lateral malleoli, tibial tuberosities, medial and lateral femoral condyles, thigh, anterior and posterior superior iliac spines, wrists, forearms, lateral elbow condyles, triceps, acromions; an offset marker was placed over the right scapula) [20]. A static calibration trial was collected to generate each runner's inertial model (Cortex, Motion Analysis Corp, Santa Rosa, CA, USA). Kinematic data and spatial-temporal parameters, including vertical displacement of the center of mass (COM), were processed using Visual 3D software (C-Motion Inc., Rockville, MD, USA). Runners ran at a self-selected velocity defined as a pace "used for the typical long run distance." After an eight-minute acclimation running period, high-speed reference videos (300 fps) were captured in the sagittal and frontal planes, and a 10-s sample of data was captured, including an average of 12–14 strides. These reference videos were captured and reviewed with each participant later to provide feedback on the subtleties of their kinematic gait motion features at various points in the gait cycle. Videos captured in the sagittal and frontal planes (Casio Elixim EX-FH20; Casio America, Inc., Dover, NJ, USA).

Kinetic Data Preprocessing. We apply here a recently applied methodological approach that serves to capture and preserve the most representative GRF signals possible [21]. The resultant magnitude of the net GRF was calculated as the square root of the sum of the squares of all three-dimensional force components without filtering the raw signal of the instrument. Net GRFs were normalized by stance time, averaging 3–4 samples for each percent of stance to preserve instantaneous rate-change information and signal fidelity across all steps for the left and right limb foot strike waveforms. Loading and unloading were characterized by LRs and calculated for every percentage of stance using a central difference numerical derivative of the normalized net GRF with the time interval equal to 1/100th of the respective left and right limb average stance times. GRFs and LRs have been normalized by body weight to facilitate comparisons across runners.

GRF Waveform Modeling. The net GRF waveform exhibits bimodal features, characterized by two distinct phases: the impact phase and the active phase. Net GRF was modeled as a double-mass-spring dynamic system. This approach captures the composite effect of loading across the lower extremity during impact and the remainder of the body during the active phase. To characterize the temporal structure of the net GRF, we applied a double Gaussian curve-fitting technique, modeling the signal as the sum of two overlapping bell-shaped impulses. Each impulse was defined by its amplitude (*A*), peak time (*B*), and width or duration (*C*). The net force (*F*) as a function of time (*t*) is given by Equation (1).

$$F(t) = A_{impact}e^{-\frac{(t-B_{impact})^2}{C_{impact}^2}} + A_{active}e^{-\frac{(t-B_{active})^2}{C_{active}^2}} \quad (1)$$

This modeling approach, along with the parameters in Equation (1), enabled us to quantify and compare impulse structure across conditions, providing insight into loading distribution strategies during running in both injured and recovering populations.

Spatiotemporal Data Post-Processing. Several standard spatiotemporal, spatial, and kinematic variables were determined to produce the reference values of these measures by age bracket, and to show the performance of this sample compared with other published evidence. Bone models were developed for each runner with the individual center of mass location using commercially available software (Visual3D, C-Motion, Inc.; Germantown, MD, USA) [18,22]. Marker data were filtered at 9 Hz with a fourth-order, low-pass Butterworth filter. Bone models were created for every runner with an individual center of mass (COM) location following the methods of de Leva et al. [23]. Gait cycle time is presented in percent (0% = initial foot contact, 100% = same foot contact post-swing phase). Cadence (steps/min) and the vertical displacement of the COM (the difference between the minimal and maximal vertical height of the COM during a gait cycle) were calculated. Vertical stiffness was estimated using the following: $K_{vert} = F_{max}/\Delta y$, where F_{max} is the peak vertical force and Δy is the maximum displacement of the COM [24]. The distance between two successive placements of the same foot was defined as the stride length. The medial-lateral distance between the proximal end position of the foot at the foot strike and the proximal end position of the foot at the next contralateral foot strike was calculated as the stride width. Stance time was defined as the period when each foot was in contact with the treadmill.

Foot strike type was determined by the angle between the foot segment and the horizontal ground at foot contact, and the investigators visually confirmed this with the high-speed videos. Runners were categorized into rearfoot and non-rearfoot strikers.

2.5. Statistics

Statistical analyses were performed using SPSS version 29.0 (IBM, Armonk, NY, USA). Normality of the data (skewness and kurtosis) was evaluated using the Kolmogorov-Smirnov test, and descriptive statistics were calculated for all study variables and demographics. The assumptions of normality and analysis of variance (ANOVA) were tested on continuous demographic, anthropometric, and training history variables to determine if differences existed between diagnosis groups (controls, Unilateral TSF, Bilateral TSF, and MTSS). Chi-square tests (χ^2) were used to determine if there were differences in categorical variables among the four study groups.

Univariate analyses of covariance (ANCOVA) were used to test for differences between the biomechanical dependent variables, including kinetic parameters (Kvert), double Gaussian parameters, running velocity, and kinematic variables. The fixed, between-group factor was diagnosis group (controls, Unilateral TSF, Bilateral TSF, and MTSS). Based on published evidence that running velocity, age, and sex can affect running biomechanics [25], these variables were entered as covariates. The eta squared (η^2) values were generated to show the effect sizes for continuous variables; values of 0.01, 0.06, and 0.14 represented negligible to small, medium, and large effects [26]. Phi values (ϕ) were determined as effect sizes for categorical variables. Bivariate Pearson correlations were performed between tibial pain severity during running and months since diagnosis and key loading features, net GRF in the early-stance (impact) phase, and maximal positive and negative LR in the early stance. Regression equations were developed using the ordinary least squares method, and the residuals (differences between the observed y-values and the predicted y-values) were plotted. To assess model fit quality, we calculated the root mean square error (RMSE) as a percentage of peak net GRF. Statistical significance was established in advance at $p < 0.05$.

3. Results

3.1. Participant Characteristics

Table 1 provides the characteristics of the runners in this analysis. Overall, participants were well matched for height, weight, BMI, sex, race, and foot strike type. With respect to running-related history, there were no statistical differences in years of experience or sessions per week. More runners with UL TSF were performing speedwork at the time of injury than other groups ($p < 0.001$). Tibial pain severity during running was rated by participants to be 2.8 ± 2.2 points (0–7 point range from 0–10 points) in the MTSS group.

Table 1. Characteristics of runners without injury, with tibial stress fracture (TSF) and medial tibial stress syndrome (MTSS). Values are means ± SD or% of the group.

Variable	Not Injured (n = 33)	MTSS (n = 12)	Unilateral TSF (n = 15)	Bilateral TSF (n = 6)	p or ϕ	η^2
Age (yr)	24.9 ± 11.3	30.6 ± 15.0	21.4 ± 9.3	20.2 ± 7.2	0.16	0.08
Height (m)	1.68 ± 0.1	1.65 ± 0.1	1.69 ± 0.1	1.68 ± 0.1	0.84	0.01
Body mass (kg)	62.5 ± 11.8	64.1 ± 12.4	62.1 ± 12.9	62.1 ± 10.6	0.97	0.00
BMI (kg/m^2)	21.9 ± 2.9	23.6 ± 4.4	21.7 ± 3.3	21.9 ± 4.1	0.49	0.04
Female (#,%)	25 (62.5)	12 (60.0)	9 (60.0)	4 (80.0)	0.41	0.53

Table 1. *Cont.*

Variable	Not Injured (n = 33)	MTSS (n = 12)	Unilateral TSF (n = 15)	Bilateral TSF (n = 6)	p or ϕ	η^2
Caucasian race (#,%)	25 (75.8)	11 (91.7)	11 (73.3)	6 (100)	0.56	0.56
Duration since injury (mo)	---	6.8 ± 3.3	7.2 ± 5.0	5.0 ± 1.0	0.68	0.09
Running History						
Years (#)	6.7 ± 7.3	6.7 ± 5.7	5.1 ± 7.4	9.0 ± 5.7	0.69	0.03
Sessions (#/wk)	4.0 ± 1.6	4.1 ± 1.1	3.6 ± 1.9	3.0 ± 2.4	0.34	0.05
Volume (km/wk)	36.4 ± 20.4	27.0 ± 15.9	34.1 ± 23.3	16.8 ± 17.4	0.22	0.07
Doing speedwork (#,%)	6 (18.2)	7 (58.3)	11 (73.3)	2 (33.3)	<0.001	0.69
Foot strike type (#,%)						
Rearfoot	28 (84.8)	11 (91.7)	10 (66.7)	6 (100.0)		
Non-rearfoot	5 (15.2)	1 (8.3)	5 (33.3)	0 (0.0)	0.49	0.36

BMI = body mass index; # = count.

3.2. Kinetic Parameters and K_{vert}

Summary statistics for key kinetic parameters and K_{vert} are presented in Table 2. Peak medial GRFs were 18–43% higher in the BL TSF group compared to MTSS and UL TSF (both $p < 0.05$). Peak anterior GRF was highest in the BL TSF group ($p = 0.02$). The effect sizes for these parameters were large (η^2 value range = 0.14–0.16). Peak vertical GRF values for all four diagnosis groups occurred during the active phase of stance (45% of stance). Peak medial and lateral GRFs occurred during the impact phase (<20% of stance). Peak anterior and posterior GRF for all four diagnosis groups occurred at 23% and 75%, respectively, during the active phase of stance. RMSE values for the net GRF Gaussian model across all groups were low relative to peak net force (left side RMSE ranged from 3.1% to 3.8% and right side RMSE ranged from 2.7% to 4.1%). For the right-sided RMSE, MTSS had the lowest value, and BL TSF had the highest value.

Figure 1A–C show the GRF waveforms during stance for all four study groups. Panel 1A reveals that the healthy control runners and BL TSF produced higher peak impact net GRF than UL TSF and MTSS groups ($h^2 = 0.145$; $p < 0.05$). Panel 1B shows that the BL TSF group had higher peak medial and lateral GRFs than the remaining groups ($p < 0.05$). The η^2 values ranged from 0.156–0.199 for medial forces and 0.064–0.156 for lateral forces. There were no group differences with respect to AP forces (Panel 1C).

Figure 2 presents the LR waveforms during stance and indicates that the BL TSF and healthy controls had higher peak positive LR values than the UL TSF and MTSS (102.2 ± 35.8 BW/s and 104.2 ± 43.7 BW/s versus 89.0 ± 31.8 BW/s and 82.1 ± 41.4 BW/s, respectively). The greatest minimum LR occurred in healthy controls and MTSS compared to UL TSF and BL TSF (−23.2 ± 27.6 BW/s and −23.8 ± 23.7 BW/s versus −8.0 ± 30.0 BW/s and −11.8 ± 23.1 BW/s, respectively). The η^2 values ranged from 0.02–0.07 for maximum LR and 0.06–0.07 for minimum LR, and were considered small. Impact and active phase impulses, braking and propulsion impulses, and medial and lateral impulses are shown in Supplementary Figure S1. While there were no statistical differences in the mean impulse values among the four groups, the effect size range was considered small to medium. The η^2 values were the following for each impulse dimension: impact = 0.032, active = 0.067, braking = 0.030, propulsion = 0.017, medial = 0.067, and lateral = 0.062.

Table 2. Summary statistics for kinetic parameters and leg stiffness (K_{vert}) among runners without injury, with unilateral or bilateral tibial stress fracture (UL TSF or BL TSF), and medial tibial stress syndrome (MTSS). Values are means ± SD, covaried for age, sex, and running velocity.

Variable	Not Injured (*n* = 33)	MTSS (*n* = 12)	UL TSF (*n* = 15)	BL TSF (*n* = 6)	*p*	η^2
GRF components (BW)						
Peak vertical GRF						
Left	2.58 ± 0.50	2.26 ± 0.32	2.39 ± 0.20	2.34 ± 0.17	0.19	0.08
Right	2.57 ± 0.49	2.23 ± 0.34	2.43 ± 0.23	2.41 ± 0.29	0.24	0.07
Peak anterior GRF						
Left	0.13 ± 0.02	0.13 ± 0.04	0.13 ± 0.02	0.14 ± 0.04	0.52	0.04
Right	0.12 ± 0.04	0.14 ± 0.06	0.13 ± 0.03	0.19 ± 0.09 *	0.02	0.16
Peak posterior GRF						
Left	−0.12 ± 0.08	−0.12 ± 0.03	−0.14 ± 0.04	−0.12 ± 0.03	0.85	0.01
Right	−0.12 ± 0.09	−0.12 ± 0.04	−0.12 ± 0.03	−0.12 ± 0.04	0.99	0.01
Peak medial GRF						
Left	0.11 ± 0.04	0.09 ± 0.05	0.13 ± 0.06	0.16 ± 0.09 *	0.03	0.14
Right	0.16 ± 0.07	0.10 ± 0.04	0.13 ± 0.05	0.16 ± 0.10 **	0.01	0.16
Peak lateral GRF						
Left	0.12 ± 0.09	0.07 ± 0.04	0.05 ± 0.05	0.06 ± 0.06	0.10	0.10
Right	0.10 ± 0.07	0.07 ± 0.05	0.05 ± 0.03	0.08 ± 0.07	0.21	0.07
RMSE net GRF (BW)						
Left	0.09 ± 0.03	0.07 ± 0.01	0.09 ± 0.03	0.09 ± 0.02	0.15	0.09
Right	0.09 ± 0.03	0.06 ± 0.02 *	0.09 ± 0.03	0.10 ±0.03	0.02	0.14
Load Rate (LR; BW/s)						
Maximum LR						
Left	104.2 ± 43.2	87.1 ± 27.4	87.5 ± 43.9	94.2 ± 31.1	0.79	0.02
Right	104.2 ± 44.3	90.9 ± 36.3	76.7 ± 38.9	110.1 ± 40.6	0.32	0.06
Minimum LR						
Left	−24.2 ± 30.1	−20.9 ± 22.5	−8.4 ± 29.7	−10.2 ± 21.9	0.31	0.06
Right	−22.2 ± 25.1	−26.7 ± 24.8	−7.5 ± 30.2	−13.3 ± 24.2	0.23	0.07
K_{vert} (N/cm)	173 ± 34	165 ± 35	169 ± 42	164 ± 45	0.47	0.04

* different than all other groups at $p < 0.05$ ** different than Unilateral TSF, MTSS at $p < 0.05$.

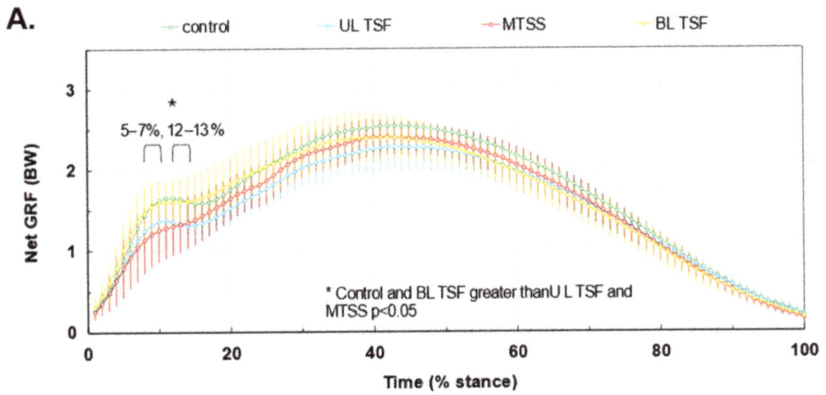

Figure 1. *Cont.*

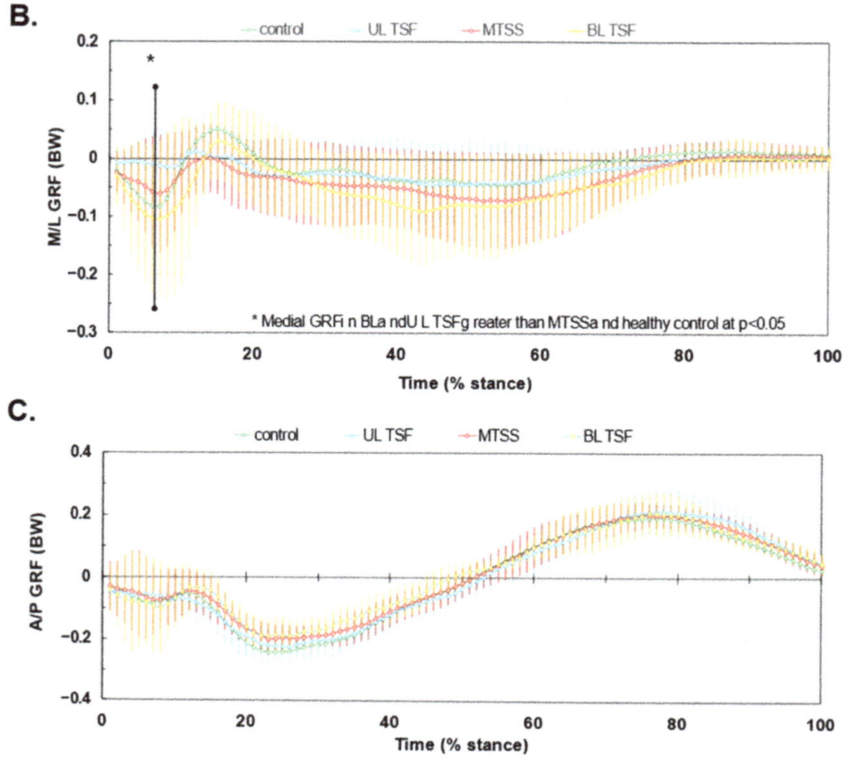

Figure 1. GRF waveforms during stance for tibial stress injuries. Panel (**A**): net GRF. Panel (**B**): Medial-lateral (M/L) GRF with negative being medially directed. Panel (**C**) Anterior-posterior (A/P) force components, with positive being propulsive GRFs. Waveforms are normalized to body weight and stance duration. Values are means ± SD. * denotes significant difference among groups.

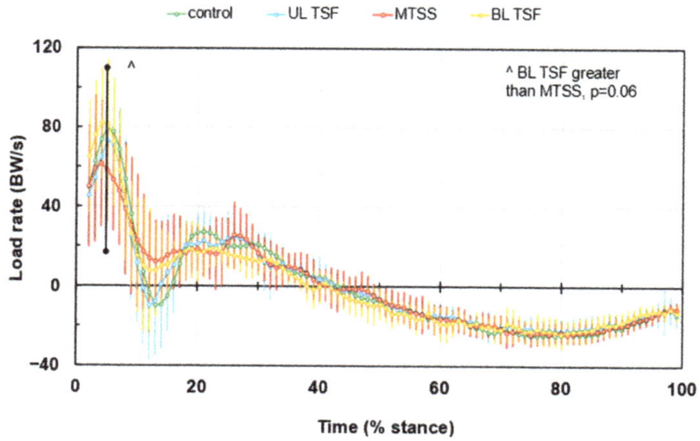

Figure 2. Load rates (LR) during an average gait cycle. Values are means ± SD. ^ denotes group difference trend at $p = 0.06$.

3.3. Gaussian Parameters

Table 3 provides a summary of the Gaussian parameters during the impact phase and active phase of loading during stance. A significant difference emerged for Impact Phase (A), where the highest impact force value differed by group for the right limb ($p = 0.02$).

Table 3. Summary statistics of double-Gaussian parameters for net GRF waveforms during the impact and active phases of stance among runners with no injury and post-recovery from UL TSF or BL TSF, and symptomatic MTSS. Note: Values are means ± SD, covariates are age, sex, and running velocity. Different superscripts indicate significant group differences at $p < 0.05$. Net GRF is the resultant magnitude, including all planar force components. Waveforms were modeled with a double-Gaussian, where A represents peak amplitude (normalized to body weight), B the peak time (% stance), and C the duration (ms) of each phase.

Variable		Not Injured ($n = 33$)	MTSS ($n = 12$)	UL TSF ($n = 15$)	BL TSF ($n = 6$)	p	η^2
Impact phase							
A (BW)	Left	0.75 ± 0.41	0.74 ± 0.38	0.65 ± 0.47	0.77 ± 0.67	0.72	0.02
	Right	0.70 ± 0.34	0.81 ± 0.56 *	0.51 ± 0.45	0.61 ± 0.36	0.02 *	0.14
B (ms)	Left	27.2 ± 10.0	30.5 ± 16.2	28.8 ± 16.6	30.4 ± 11.2	0.57	0.03
	Right	26.9 ± 9.8	30.3 ± 17.5	26.6 ± 10.0	26.8 ± 7.4	0.79	0.02
C (ms)	Left	12.9 ± 11.6	15.8 ± 15.9	17.0 ± 17.3	18.1 ± 14.3	0.33	0.06
	Right	13.7 ± 12.5	15.0 ± 15.2	12.5 ± 12.5	13.4 ± 6.4	0.95	0.01
Active phase							
A (BW)	Left	2.58 ± 0.50	2.26 ± 0.32	2.39 ± 0.20	2.34 ± 0.17	0.19	0.08
	Right	2.57 ± 0.50	2.23 ± 0.34	2.43 ± 0.22	2.41 ± 0.29	0.24	0.07
B (ms)	Left	111.2 ± 13.4	116.4 ± 23.7	115.5 ± 18.8	108.5 ± 18.8	0.84	0.01
	Right	112.1 ± 14.1	118.8 ± 22.7	114.5 ± 14.3	104.9 ± 21.4	0.75	0.02
C (ms)	Left	87.8 ± 11.2	88.1 ± 17.6	85.2 ± 13.2	85.8 ± 17.2	0.35	0.05
	Right	88.4 ± 10.9	88.3 ± 17.9	86.9 ± 13.4	90.5 ± 10.9	0.53	0.04

Impact A= impact peak amplitude; B = impact peak time; C = impact peak duration. Active A = active peak amplitude; B = active peak time; C = active peak duration; * different than all other groups at $p < 0.05$.

3.4. Scatterplots

Figure 3 provides the scatter plots with linear regressions of forces and LRs by tibial pain severity (0–10 points) and duration since diagnosis in the impact phase (1–20% stance). Bivariate correlation coefficients are provided in the figure inset. Panels A, C, and E indicate that peak GRF and LR are inversely correlated to pain severity, and peak negative LR is positively correlated to pain during the impact phase. Panels D and F indicate that peak positive and negative LRs are inversely and directly related to duration since diagnosis. Correlations in Figure 3A,C–E were significant at $p < 0.05$.

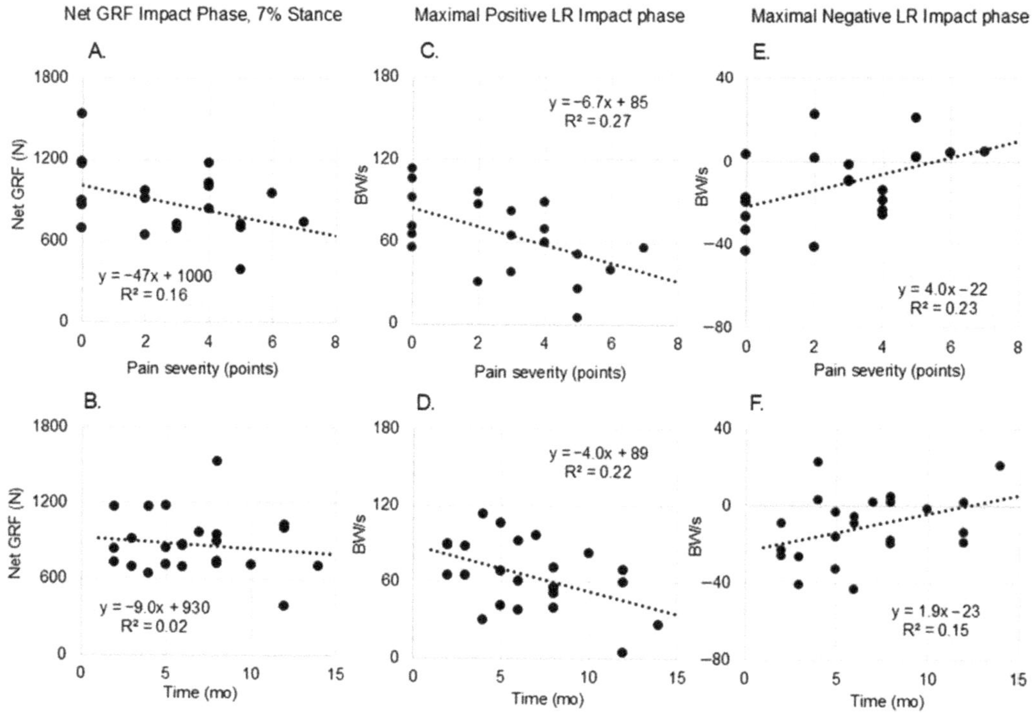

Figure 3. Scatter plots of impact phase net ground reaction force (GRF) values and maximal positive and negative load rates (LR) by tibial pain severity (0–10 points) and by duration since tibial bone injury diagnosis. Panels (**A**,**B**): Net GRF in the impact phase of stance (7%) as a function of pain ((**A**); r = 0.512) and duration since diagnosis ((**B**); r = 0.130); Panels (**C**,**D**): Maximal positive LR during the impact phase as a function of pain ((**C**); r = 0.522) and duration since diagnosis ((**D**); r = 0.464); and Panels (**E**,**F**): Maximal negative LR during the impact phase as a function of pain ((**E**); r = 0.481) and duration since diagnosis ((**F**); r = 0.389). All correlation values for these scatterplots, except panel (**B**), were significant at $p < 0.05$.

3.5. Velocity and Spatiotemporal Parameters

Supplementary Table S1 provides details about running velocity and select spatiotemporal parameters. None of these measures (velocity, step lengths, stride width, COM vertical displacement, stance times) were found to be statistically different by study group.

4. Discussion

This study characterized the multidimensional GRF features among runners with different acuities of tibial stress injury compared to healthy non-injured controls. By applying a novel data analytic approach to force data using Gaussian modeling, we found several nuances to the GRF response by injury acuity, largely early in the stance phase. The novel findings from this study demonstrate that runners with BL TSF generate higher peak net GRF and LR in early stance compared to runners with UL TSF and MTSS. Additionally, BL TSF runners produced elevated mediolateral GRFs relative to other groups. Importantly, maximal positive and negative LR were differentially related to pain severity and time since diagnosis, highlighting the value of integrating symptom severity and injury acuity into biomechanical analyses to contextualize gait results for each patient. The techniques presented here may offer investigators a new approach to disentangling the roles of GRF and LR in specific running-related injuries.

To our knowledge, these are among the first data to characterize running biomechanics and loading among individuals with recent bilateral tibial TSF compared to UL TSF and MTSS. Our results show that runners recovering from BL TSF exhibit relatively higher net GRF (Figure 1A), altered mediolateral force distributions (Figure 1B), and positive LR (Figure 2) than the remaining groups. These patterns existed despite very similar temporal spatial parameters and running velocities (Supplementary Table S1). Bilateral tibial injury likely induces more global gait adaptations compared to unilateral injury or MTSS, as one limb may not be used as a compensatory side during running—both sides must accept the loading of body weight. While the bone may be healed and pain is no longer present, the nervous system might still be adapting to the changes in movement and weight-bearing, potentially leading to persistent gait alterations [27]. Available evidence to support this concept is from surgical patients who experienced tibial fractures and repairs [27]. Despite full bone healing, lower GRF peaks existed among patients with tibial shaft fractures compared to healthy controls by 6 months or longer, which is paralleled by chronic impairments in squat motion [28]. Function and gait often normalize by 12 months or beyond [27]. Our runners tested within the first 4–12 months post-diagnosis, so it is reasonable to find that GRF and LR varied from other groups during testing in this same post-injury time window.

Runners with symptomatic MTSS were characterized by the smallest LR and medial GRF during the first 20% of stance, whereas the recovered BL TSF produced the highest medial GRF during this time (Figure 1B). These results contradicted our hypothesis. Peak net GRF and maximal positive LR were inversely associated with pain severity and time since diagnosis, while maximal negative LR was positively associated with both. Runners experiencing higher levels of pain, and those closer to the time of diagnosis, unloaded more during the impact phase of stance (Figure 3); these runners may be using a self-imposed unloading strategy due to pain avoidance or represent early adaptations to emerging bone stress. Among runners who have muscle soreness, GRF values during the days after onset are reduced commensurate with elevated GRF variance [29]. Pain is related to kinesiophobia (fear or movement due to pain). Specifically, pain can interrupt normal movement and exposure to pain and drive a new learned motor pathway to prepare for the impending 'threat' of pain [30] or elevated [31]. One possibility is that, based on the tibial location where the pain is occurring (e.g., medial or lateral tibial shaft, proximal or distal), the runner may try to offload the specific site of injury by adjusting muscle activation patterns and shifting medial-lateral GRFs. For runners with MTSS and injury along the anterior spine, reducing large LR excursions and peak GRF overall may help offload this relatively large painful area without a medial or lateral GRF shift. Among individuals with BL TSF who are returning to running and trying to prevent reinjury, controlling loading and LR may be especially challenging. Large medial GRF with large variance may manifest as shown in the present study. With high variability but a relatively lower average pain level (2.8 out of 10 points), the MTSS group had some similarities with the uninjured running group. However, on several key metrics, they demonstrated that they made accommodations associated with higher pain severities, most clearly in Figure 2, including an average lower maximum LR and higher minimum LR in the impact phase. MTSS is acute, resulting in high variability in runners as they react to relatively new and unfamiliar stimuli. However, it is a less severe injury, and it appears that runners can intuitively optimize specific parameters to reduce the pain associated with their running mechanics.

Additional larger cohorts of the BL TSF subgroup combined with either electromyography or electroencephalopathy could help confirm this finding and clarify central and peripheral mechanisms relating to early stance loading.

Our findings differ from others that compared LRs in early stance between healthy runners and a variety of injured runners grouped together or by individual injuries (TSF, Achilles tendinopathy, patellofemoral pain, Iliotibial band pain) [32]. In this previous study, TSF injuries were not related to either vertical or medial-lateral LR. Other evidence on this relationship is mixed, with some studies reporting associations of TSF and LR [33,34] whereas others do not. Our study provided GRF loading responses across the spectrum of tibial bone injury, which provided insight into how loading may be involved in this injury type. From a clinical perspective, persistent mechanical loading imbalances and compensation strategies are observed after returning to running. Emphasis return-to-run progression, particularly in assessing readiness and symmetry of loading. The findings also emphasize that MTSS may represent a distinct biomechanical profile from TSF, possibly requiring unique therapeutic strategies targeting different phases of the gait cycle. Moreover, the use of alternative preprocessing methods that avoid heavy signal filtering proved essential in detecting these nuanced waveform features. By preserving high-frequency kinetic information, our approach facilitated the identification of subtle but potentially clinically relevant signal characteristics, including biphasic LR fluctuations and mediolateral GRF shifts.

Limitations and Future Directions: Several limitations must be acknowledged. First, although we matched healthy controls on key variables (e.g., sex, BMI, experience), residual confounding due to unmeasured variables, such as neuromuscular control or psychological factors, may still remain. Second, the cross-sectional design prohibits causal inference regarding whether the observed biomechanical differences are precursors to injury or consequences of compensatory adaptation. Longitudinal follow-up of runners at a baseline time point, through injury onset and recovery, is necessary to clarify these relationships. Third, while we controlled for footwear and training history, individual differences in surface compliance, running technique, or strength training practices could also influence GRF characteristics. Lastly, the number of healthy controls engaging in speedwork during training was significantly lower than that of the injured groups. Previous work has shown that higher running speeds are associated with higher tibial strain measurements [35]. Data collection in these runners occurred at a self-selected, comfortable running speed; thus, it is possible that our data does not fully account for kinetic alterations at higher speeds. Future research could expand on these findings by integrating muscle activation patterns, electroencephalography, and repeated testing of runners during the injury-healing process to track kinetic changes over time. Prospective studies using similar signal-preserving preprocessing methods described here may help identify whether specific waveform features predict injury risk or recurrence.

5. Conclusions

This study provides novel insight into the biomechanical patterns associated with tibial bone stress injury and recovery in runners. By examining high-resolution GRF and LR waveforms using signal-preserving modeling, we revealed key differences among runners with MTSS, UL TSF, BL TSF, and healthy controls. Our findings indicate that BL TSF runners exhibit higher medial GRF and LR values, potentially reflecting more global mechanical adaptations, whereas runners with MTSS demonstrate unloading in early stance. Pain severity and time since diagnosis were significantly related to GRF and LR features, highlighting the importance of considering symptom context when interpreting biomechanical data. These findings may inform more tailored rehabilitation strategies and support the utility of advanced signal analysis methods in running biomechanics research.

Supplementary Materials: The following supporting information can be downloaded at: https://www.mdpi.com/article/10.3390/bioengineering12080802/s1, Figure S1: Impulses during stance phase of running among runners without injury, with tibial stress fracture (TSF) and medial tibial stress syndrome (MTSS). Table S1: Running velocity and key spatiotemporal parameters among runners without injury, with tibial stress fracture (TSF) and medial tibial stress syndrome (MTSS).

Author Contributions: Conceptualization, R.M.N. and H.K.V.; methodology, R.M.N., H.K.V. and S.S.; validation, R.M.N. and H.K.V.; formal analysis, H.K.V., S.S. and M.M.; investigation, R.M.N., S.S., M.M., L.P., K.R.V. and H.K.V.; resources, H.K.V. and K.R.V.; data curation, H.K.V., S.S., M.M., L.P., and R.M.N.; writing—original draft preparation, All authors; writing—review and editing, All Authors; visualization, R.M.N., K.R.V. and H.K.V.; supervision, R.M.N. and H.K.V.; project administration, L.P., S.S. and H.K.V.; funding acquisition, H.K.V. and K.R.V. All authors have read and agreed to the published version of the manuscript.

Funding: This research was supported in part by the UF Strategic Funding Initiative, "Transforming Sport Science for Every Body" (PIs; H and K Vincent).

Institutional Review Board Statement: This secondary analysis study was conducted in accordance with the Declaration of Helsinki, and approved by the Institutional Review Board of the University of Florida (protocol IRB # 202500639 4.21.2025)." for studies involving humans.

Informed Consent Statement: Patient consent was waived due to the fact that the methods involved previously collected data in our d approved departmental databank, and the review board approved the study waiving consent.

Data Availability Statement: The raw data supporting the conclusions of this article could be made available by the authors upon reasonable request and review by the study team. This is also contingent upon agreement with the University of Florida and appropriate data use agreements.

Conflicts of Interest: Dr. Heather Vincent currently serves on the Board of Trustees of the American College of Sports Medicine.

Abbreviations

The following abbreviations are used in this manuscript:

MTSS	Medial tibial stress syndrome
TSF	Tibial stress fracture
UL	unilateral
BL	bilateral
GRF	ground reaction force
LR	load rate
COM	center of mass
BMI	body mass index

References

1. Lopes, A.D.; Hespanhol Júnior, L.C.; Yeung, S.S.; Costa, L.O.P. What Are the Main Running-Related Musculoskeletal Injuries? A Systematic Review. *Sports Med. Auckl. NZ* **2012**, *42*, 891–905. [CrossRef]
2. Bhusari, N.; Deshmukh, M. Shin Splint: A Review. *Cureus* **2023**, *15*, e33905. [CrossRef]
3. Galbraith, R.M.; Lavallee, M.E. Medial Tibial Stress Syndrome: Conservative Treatment Options. *Curr. Rev. Musculoskelet. Med.* **2009**, *2*, 127–133. [CrossRef]
4. Behrens, S.B.; Deren, M.E.; Matson, A.; Fadale, P.D.; Monchik, K.O. Stress Fractures of the Pelvis and Legs in Athletes: A Review. *Sports Health* **2013**, *5*, 165–174. [CrossRef]
5. Matheson, G.O.; Clement, D.B.; McKenzie, D.C.; Taunton, J.E.; Lloyd-Smith, D.R.; MacIntyre, J.G. Stress Fractures in Athletes. A Study of 320 Cases. *Am. J. Sports Med.* **1987**, *15*, 46–58. [CrossRef]
6. Pohl, M.B.; Mullineaux, D.R.; Milner, C.E.; Hamill, J.; Davis, I.S. Biomechanical Predictors of Retrospective Tibial Stress Fractures in Runners. *J. Biomech.* **2008**, *41*, 1160–1165. [CrossRef]
7. Yagi, S.; Muneta, T.; Sekiya, I. Incidence and Risk Factors for Medial Tibial Stress Syndrome and Tibial Stress Fracture in High School Runners. *Knee Surg. Sports Traumatol. Arthrosc. Off. J. ESSKA* **2013**, *21*, 556–563. [CrossRef]

8. Komaris, D.-S.; Perez-Valero, E.; Jordan, L.; Barton, J.; Hennessy, L.; O'Flynn, B.; Tedesco, S. Effects of Segment Masses and Cut-off Frequencies on the Estimation of Vertical Ground Reaction Forces in Running. *J. Biomech.* **2020**, *99*, 109552. [CrossRef]
9. Zadpoor, A.A.; Nikooyan, A.A. The Relationship between Lower-Extremity Stress Fractures and the Ground Reaction Force: A Systematic Review. *Clin. Biomech. Bristol Avon* **2011**, *26*, 23–28. [CrossRef]
10. Gruber, A.H. The "Impacts Cause Injury" Hypothesis: Running in Circles or Making New Strides? *J. Biomech.* **2023**, *156*, 111694. [CrossRef]
11. Mai, P.; Willwacher, S. Effects of Low-Pass Filter Combinations on Lower Extremity Joint Moments in Distance Running. *J. Biomech.* **2019**, *95*, 109311. [CrossRef]
12. Bramah, C.; Preece, S.J.; Gill, N.; Herrington, L. Is There a Pathological Gait Associated with Common Soft Tissue Running Injuries? *Am. J. Sports Med.* **2018**, *46*, 3023–3031. [CrossRef]
13. Pantak, M. Ground Reaction Forces Generated by Runners-Harmonic Analyses and Modelling. *Appl. Sci.* **2020**, *10*, 1575. [CrossRef]
14. Messier, S.P.; Martin, D.F.; Mihalko, S.L.; Ip, E.; DeVita, P.; Cannon, D.W.; Love, M.; Beringer, D.; Saldana, S.; Fellin, R.E.; et al. A 2-Year Prospective Cohort Study of Overuse Running Injuries: The Runners and Injury Longitudinal Study (TRAILS). *Am. J. Sports Med.* **2018**, *46*, 2211–2221. [CrossRef]
15. Riley, P.O.; Dicharry, J.; Franz, J.; Della Croce, U.; Wilder, R.P.; Kerrigan, D.C. A Kinematics and Kinetic Comparison of Overground and Treadmill Running. *Med. Sci. Sports Exerc.* **2008**, *40*, 1093–1100. [CrossRef]
16. Vincent, H.K.; Popp, R.; Cicilioni, O.; Vincent, K.R.; Pezzullo, L.; Martenson, M.; Nixon, R.M. Reference Biomechanical Parameters and Natural Asymmetry among Runners across the Age Spectrum without a History of Running-Related Injuries. *Front. Sports Act. Living* **2025**, *7*, 1560756. [CrossRef]
17. Vincent, H.K.; Vincent, K.R. Evaluation of the Injured Runner. In *Clinical Care of the Runner: Assessment, Biomechanical Principles, and Injury Management*; Elsevier: Amsterdam, The Netherlands, 2020; pp. 19–26.
18. Madsen, A.; Sharififar, S.; Oberhaus, J.; Vincent, K.R.; Vincent, H.K. Anxiety State Impact on Recovery of Runners with Lower Extremity Injuries. *PLoS ONE* **2022**, *17*, e0278444. [CrossRef]
19. Vincent, H.K.; Massengill, C.; Harris, A.; Chen, C.; Wasser, J.G.; Bruner, M.; Vincent, K.R. Cadence Impact on Cardiopulmonary, Metabolic and Biomechanical Loading during Downhill Running. *Gait Posture* **2019**, *71*, 186–191. [CrossRef]
20. Kadaba, M.P.; Ramakrishnan, H.K.; Wootten, M.E. Measurement of Lower Extremity Kinematics during Level Walking. *J. Orthop. Res. Off. Publ. Orthop. Res. Soc.* **1990**, *8*, 383–392. [CrossRef]
21. Nixon, R.M.; Vincent, K.R.; Pezzullo, L.; Vincent, H.K. Enhanced Ground Reaction Force Analyses Reveal Injury-Related Biomechanical Differences in Runners. *Sci. Rep. in review*. 2025.
22. Vincent, H.K.; Kilgore, J.E.; Chen, C.; Bruner, M.; Horodyski, M.; Vincent, K.R. Impact of Body Mass Index on Biomechanics of Recreational Runners. *PM R* **2020**, *12*, 1106–1112. [CrossRef]
23. de Leva, P. Joint Center Longitudinal Positions Computed from a Selected Subset of Chandler's Data. *J. Biomech.* **1996**, *29*, 1231–1233. [CrossRef]
24. McMahon, T.A.; Cheng, G.C. The Mechanics of Running: How Does Stiffness Couple with Speed? *J. Biomech.* **1990**, *23* (Suppl. 1), 65–78. [CrossRef]
25. Furlong, L.-A.M.; Egginton, N.L. Kinetic Asymmetry during Running at Preferred and Nonpreferred Speeds. *Med. Sci. Sports Exerc.* **2018**, *50*, 1241–1248. [CrossRef]
26. Cohen, J. *Statistical Power Analysis for the Behavioral Sciences*; Routledge: New York, NY, USA, 1988.
27. Warmerdam, E.; Orth, M.; Pohlemann, T.; Ganse, B. Gait Analysis to Monitor Fracture Healing of the Lower Leg. *Bioengineering* **2023**, *10*, 255. [CrossRef]
28. Kröger, I.; Müßig, J.; Brand, A.; Pätzold, R.; Wackerle, H.; Klöpfer-Krämer, I.; Augat, P. Recovery of Gait and Function during the First Six Months after Tibial Shaft Fractures. *Gait Posture* **2022**, *91*, 66–72. [CrossRef]
29. Markus, I.; Arutiunian, A.; Ohayon, E.; Holodov, M.; Peled, D.; Yavetz, R.; Ben-Eliezer, N.; Dello Iacono, A.; Gepner, Y. Kinetics of Recovery and Normalization of Running Biomechanics Following Aerobic-Based Exercise-Induced Muscle Damage in Recreational Male Runners. *J. Sci. Med. Sport* **2025**, *28*, 498–505. [CrossRef]
30. Ohji, S.; Aizawa, J.; Hirohata, K.; Ohmi, T.; Mitomo, S.; Koga, H.; Yagishita, K. Association between Landing Biomechanics, Knee Pain, and Kinesiophobia in Athletes Following Anterior Cruciate Ligament Reconstruction: A Cross-Sectional Study. *PM R* **2023**, *15*, 552–562. [CrossRef]
31. Trigsted, S.M.; Cook, D.B.; Pickett, K.A.; Cadmus-Bertram, L.; Dunn, W.R.; Bell, D.R. Greater Fear of Reinjury Is Related to Stiffened Jump-Landing Biomechanics and Muscle Activation in Women after ACL Reconstruction. *Knee Surg. Sports Traumatol. Arthrosc. Off. J. ESSKA* **2018**, *26*, 3682–3689. [CrossRef]
32. Johnson, C.D.; Tenforde, A.S.; Outerleys, J.; Reilly, J.; Davis, I.S. Impact-Related Ground Reaction Forces Are More Strongly Associated with Some Running Injuries than Others. *Am. J. Sports Med.* **2020**, *48*, 3072–3080. [CrossRef]

33. Zifchock, R.A.; Davis, I.; Hamill, J. Kinetic Asymmetry in Female Runners with and without Retrospective Tibial Stress Fractures. *J. Biomech.* **2006**, *39*, 2792–2797. [CrossRef]
34. Milner, C.E.; Foch, E.; Gonzales, J.M.; Petersen, D. Biomechanics Associated with Tibial Stress Fracture in Runners: A Systematic Review and Meta-Analysis. *J. Sport Health Sci.* **2023**, *12*, 333–342. [CrossRef]
35. Baggaley, M.; Haider, I.; Bruce, O.; Khassetarash, A.; Edwards, W.B. Tibial Strains Are Sensitive to Speed Perturbations, but Not Grade Perturbations, during Running. *J. Exp. Biol.* **2024**, *227*, jeb246770. [CrossRef]

Disclaimer/Publisher's Note: The statements, opinions and data contained in all publications are solely those of the individual author(s) and contributor(s) and not of MDPI and/or the editor(s). MDPI and/or the editor(s) disclaim responsibility for any injury to people or property resulting from any ideas, methods, instructions or products referred to in the content.

bioengineering

MDPI

Review

Machine Learning Methods in Posture-Related Applications in Children up to 12 Years Old: A Systematic Review

Markel Rico-González [1,2], Carlos D. Gómez-Carmona [2,3,4,*], Ibrahim Ouergui [5,6] and Luca Paolo Ardigò [7,*]

1 Department of Didactics of Music, Plastic and Body Expression, University of Basque Country (UPV-EHU), 48940 Leioa, Spain; markel.rico@ehu.eus
2 BioVetMed & SportSci Research Group, University of Murcia, 30001 Murcia, Spain
3 Research Group in Training, Physical Activity and Sports Performance (ENFYRED), Department of Music, Plastic and Body Expression, University of Zaragoza, 44003 Teruel, Spain
4 Research Group in Training Optimization and Sports Performance (GOERD), University of Extremadura, 10071 Caceres, Spain
5 High Institute of Sport and Physical Education of Kef, University of Jendouba, Kef 7100, Tunisia; brahim.ouerghi@issepkef.u-jendouba.tn
6 Sport Sciences, Health and Movement, UR22JS01, University of Jendouba, Kef 7100, Tunisia
7 Department of Teacher Education, NLA University College, 0166 Oslo, Norway
* Correspondence: carlosdavid.gomez@unizar.es (C.D.G.-C.); luca.ardigo@nla.no (L.P.A.)

Abstract

One of the most important factors in how infants and young children learn to move is postural control. This systematic review aims to evaluate the machine learning methods in posture-related applications for children aged 0–12. Following PRISMA guidelines, we systematically searched the PubMed, Web of Sciences, SCOPUS, and ProQuest Central databases. Twenty-two studies were included in the qualitative synthesis following screening of 199 articles, with methodological quality assessed as moderate to good using the MINORS scale (scores ranging from 8/16 to 19/24). The reviewed research involved diverse samples of infants and children up to 12 years old, employing sensor-based technologies such as inertial measurement units, force plates, pressure mats, and video cameras to extract kinematic and postural features for machine learning applications. Reported accuracies, typically exceeding 85%, reflected considerable methodological heterogeneity related to sensor modality, data quality, and model architecture. Algorithms such as Random Forest, SVM, and CNN were most frequently and effectively applied for posture classification, early detection of developmental delays, and diagnosis of conditions such as cerebral palsy and autism spectrum disorder, demonstrating promising potential for at-home monitoring and clinical interventions.

Keywords: technology; machine learning; prediction; computer science; health

1. Introduction

Postural control is considered one of the critical aspects that designs how newborns and children learn to move, supporting balance, coordination, and interaction with their environment [1]. This can be achieved by the adaptation of different systems such as neuromuscular, sensory, and skeletal ones and their capacity to adapt quickly to both internal growth and external stress across the first months and years of life, inducing rapid changes in body alignment and movement patterns [2,3]. Correct posture in early life is crucial not only for the health of the musculoskeletal system and movement efficiency, but also for the development of motor learning and the independence of functional capacity [4,5].

However, early postural deviations—such as forward head posture, trunk asymmetry, or excessive kyphosis or scoliosis—can lead to compensation in terms of movement patterns, functional limitations, and increased risk of pain or musculoskeletal disorders later in life [6].

Visual observation, observational milestones, goniometric measurement, and photographic or 2D video analysis are commonly used to assess posture in young children in various settings [7,8]. Such methods provide valuable information, but they may be subjective, limited in temporal and spatial sampling, and may require specialized equipment or settings [9]. High-precision laboratory tools such as motion capture systems and force plates can result in accurate, reliable, and rich kinematic and kinetic data, but are resource-intensive and not feasible for routine clinical use or home-based monitoring due to their high cost, specialized personnel requirements, and controlled environment needs, with young children often unable to adapt well to laboratory settings [10,11]. This gap between the need for early, frequent, and ecologically valid assessment and the limitations of current methods has spurred interest in more scalable, objective, and technology-driven alternatives [9,12].

Recent advancements in machine learning (ML)—a subset of artificial intelligence (AI) that empowers computers to identify, categorize, and predict complex patterns in data— have revolutionized methodologies for posture and movement analysis [13–15]. ML models use finite details, learn how things are related in a non-linear manner, and make predictions from massive amounts of sensor, video, or image data sets [13,15,16]. While many existing ML studies have relied on laboratory-based technologies—such as wearable inertial measurement units (IMUs) to track trunk and limb movements, pressure mats, and force plates to assess postural load distribution [15–18], or 2D/depth-camera-based systems to extract keypoints and joint angles [19]—these tools mainly serve as research-grade references and remain impractical for widespread pediatric screening or routine follow-up. However, recent research showed that technological progress has enabled ML models to analyze infant postural control using accessible data sources like standard video recordings and accelerometers [1,17]. Studies have utilized computer algorithms to assess posture from images on glass platforms and evaluated infant movements through video data [1,18]. The combination of accelerometers and video has also proven its effectiveness in quantifying postural alignment in children who cannot sit independently [17]. These innovations demonstrate the efficacy of using low-cost data sources for advanced infant postural assessment in clinical and research environments. Nonetheless, this body of work demonstrates the analytic potential of ML-based posture evaluation in children, using classifiers such as Support Vector Machines (SVMs), Random Forests (RFs), and deep learning architectures like Convolutional Neural Networks (CNNs) [13,19].

Despite these advances, research specifically applying ML to pediatric posture assessment remains limited and methodologically diverse [13]. Children grow rapidly and differently from adults, with different motor patterns and unique sensor/algorithm requirements [9,20]. Additional challenges are the ethical issues (including privacy and consent), small datasets for pediatrics, and differences in how measurements are taken [9,20]. While previous reviews have addressed posture detection or ML in general, a focused synthesis of pediatric posture applications using machine learning is lacking [9,16].

Thus, this systematic review aimed to comprehensively evaluate the ML methods in posture-related applications for children aged 0–12. The specific aims were to (1) catalog sensor and vision-based technologies for pediatric posture monitoring; (2) examine ML algorithms used and performance classification; (3) clinical applications and diagnostic capabilities; (4) describe naturalistic assessment and home-based monitoring; and (5) identify methodological gaps, ethical issues, and future research directions.

2. Materials and Methods

2.1. Experimental Approach to the Problem

This systematic review followed the Preferred Reporting Items for Systematic Reviews and Meta-Analyses (PRISMA) guidelines [21] and adhered to established guidelines for performing systematic reviews in sport sciences [22]. The review protocol was developed to ensure comprehensive coverage of relevant literature while maintaining methodological rigor. The systematic review was registered (PROSPERO: CRD420251218895).

2.2. Information Sources

A comprehensive search was conducted across four databases: PubMed, Web of Sciences, SCOPUS, and ProQuest Central. The search encompassed all published literature prior to 29 August 2025. This combination of databases was selected to ensure broad coverage of both medical and sports science literature.

2.3. Search Strategy

The PICO (Patient, Problem, or Population–Intervention or Exposure–Comparison, Control, or Comparator–Outcome[s]) framework was implemented to structure the search strategy and ensure systematic coverage of relevant literature [21]. To maintain transparency, the authors were not blinded to journal names or manuscript authors. Language filters were applied to include English and Spanish publications only. The search terms were carefully selected to capture all relevant literature on information technologies and machine learning in primary school settings. The final search string was: *(child* OR infant*) AND ("machine learning") AND (postur*).*

2.4. Eligibility Criteria

The authors introduced the search string into databases and downloaded the title, author names, journal, and date of all the articles that appeared in the search. Once the Excel spreadsheet was organized, all duplicates were removed, and the remaining articles were evaluated for their eligibility (Table 1). If the authors found articles that had not appeared in the search, they included them in the Excel document as "included from external sources".

Table 1. Inclusion and exclusion criteria.

Item	Inclusion	Exclusion
Population	Children as participants (until 12 years old)	Studies with non-child participants (more than 12 years old)
Intervention or Exposure	Studies that used machine learning	Studies that did not use machine learning
Comparation	Not applicable	Not applicable
Outcome[s]	Any result (validity or reliability studies, predictions...) related to children's body posture	Any result not related to children's body posture
Other criteria	Peer-reviewed full-text studies published in original journal articles	Non-peer-reviewed journal articles Non-original full-text studies (conference papers...) Reviews

2.5. Data Extraction

A standardized data extraction process was implemented using an Excel spreadsheet developed in accordance with the Cochrane Consumers and Communication Review Group's data extraction template [23]. The spreadsheet facilitated systematic assessment of inclusion and exclusion requirements for all selected studies. Two authors independently conducted the extraction process, with any disagreements resolved through discussion until consensus was reached. Full documentation was maintained for excluded articles, including specific reasons for exclusion. All data were systematically recorded and stored in the spreadsheet.

2.6. Assessment of Study Methodology

The methodological quality was assessed using the methodological index for non-randomized studies (MINORS) [24]. The MINORS scale is a list that contains 8 essential points, and it is expanded to 12 points when the studies to be treated are comparative. In this case, it was assessed considering 9 items (out of 18 points) due to the non-possibility of applying (NA) three of them. The score that each section receives can be from 0 to 2, depending on the quality obtained by each point (0 = Low quality; 1 = Medium quality; 2 = Good quality).

3. Results

3.1. Identification and Selection of Studies

After analyzing all databases (PubMed: 27; Web of Science: 78; ProQuest Central: 14; SCOPUS: 80), the contents of 199 articles were checked; we detected, at the initial stage, 93 duplicate articles. Then, the authors analyzed whether each of the remaining 106 articles met all inclusion criteria, resulting in the elimination of 22 articles by exclusion criterion number one, exclusion criterion number two (n = 4), exclusion criterion number four (n = 12), and exclusion criterion number six (n = 46). The remaining 22 articles were included in the qualitative synthesis of the systematic review (Figure 1).

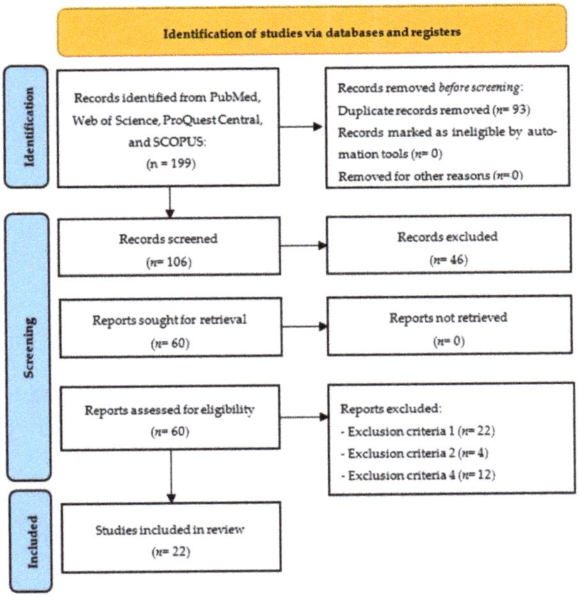

Figure 1. Flow diagram of the study.

3.2. Quality Assessment

The methodological quality of the 21 included studies was assessed using the MINORS scale, which evaluates up to 12 items (scored 0–2 each, with some non-applicable), resulting in total scores ranging from 8/16 to 19/24 (Table 2). Most studies scored good quality (2/2) on clearly defined objectives (item 1), assessments adjusted to objectives (item 4), neutral evaluations (item 5, where applicable), and appropriate statistical analysis (item 12), but consistently low quality (0/2) on prospective data collection (item 3) and sample size estimation (item 8). Overall, the studies exhibited medium to good quality, with average scores around 13/18 for non-comparative designs, indicating robust objectives and analyses but room for improvement in prospective planning and follow-up consistency.

Table 2. Methodological assessment of the included studies.

Reference	1	2	3	4	5	6	7	8	9	10	11	12	Score
Airaksinen et al. [25]	2	1	0	2	2	2	2	1	-	-	-	2	14/18
Li et al. [15]	2	1	0	2	2	2	2	2	-	-	-	2	15/18
Yang et al. [26]	2	1	1	2	2	-	-	0	-	-	-	2	10/14
Kim et al. [13]	2	1	0	2	2	2	2	0	-	-	-	2	13/18
Arias Valdivia et al. [27]	2	1	0	2	2	-	-	0	-	-	-	2	9/14
Khaksar et al. [16]	2	1	1	2	1	2	2	1	2	2	1	2	19/24
Kim et al. [14]	2	1	0	2	2	2	2	0	-	-	-	2	13/18
Lee et al. [28]	2	1	0	2	2	2	2	0	-	-	-	2	13/18
Bertoncelli et al. [29]	2	1	2	2	2	2	2	0	-	-	-	2	15/18
Sukhadia & Kamboj [30]	2	1	0	2	1	-	0	0	-	-	-	2	8/16
Li et al. [31]	2	1	0	2	2	2	-	0	-	-	-	2	11/16
Franchak et al. [32]	2	1	0	2	2	2	2	0	-	-	-	2	13/18
Eken et al. [33]	2	1	0	2	2	2	2	0	-	-	-	2	13/18
Tao et al. [34]	2	1	0	2	2	0	0	0	-	-	-	2	9/18
Ledwon et al. [35]	2	1	0	2	2	2	2	0	-	-	-	2	11/18
Airaksinen et al. [36]	2	1	0	2	2	2	2	0	-	-	-	2	13/18
Gama et al. [37]	2	1	0	2	2	-	-	0	-	-	-	2	9/14
Franchak et al. [38]	2	1	0	2	2	2	2	0	-	-	-	2	13/18
Ali & Mohamed [39]	2	0	0	2	2	2	-	0	-	-	-	2	10/16
Duda-Goławska et al. [40]	2	1	0	2	2	2	1	0	-	-	-	2	13/18
Rachwani et al. [41]	2	1	0	2	2	2	2	0	-	-	-	2	13/18

Note: The MINORS checklist: clearly defined objective (item 1); inclusion of patients consecutively (item 2); information collected retrospectively (item 3); assessments adjusted to objective (item 4); evaluations carried out in a neutral way (item 5); follow-up phase consistent with the objective (item 6); dropout rate during follow-up less than 5% (item 7); prospective estimation of sample size (item 8); adequate control group (item 9); simultaneous groups (item 10); homogeneous starting groups (item 11); and appropriate statistical analysis (item 12).

3.3. Study Characteristics

3.3.1. Sample

The included studies encompassed diverse pediatric populations, primarily infants and children up to 12 years old, with sample sizes ranging from small cohorts (e.g., 10 children in Kim et al. [13] and Eken et al. [33]) to larger groups (e.g., 514 primary school children in Tao et al. [34]). Participants included typically developing infants and children, as well as those with neurodevelopmental conditions such as autism spectrum disorder (ASD; n = 50 in Li et al. [15]), cerebral palsy (CP; up to 140 across Khaksar et al. [16] and Bertoncelli et al. [29]), HIV encephalopathy (n = 10 in Eken et al. [33]), and forward head posture (FHP; n = 514 in Tao et al. [34]). Some studies incorporated simulation models (e.g., child occupant model in Li et al. [31]) or synthetic data (e.g., MINI-RGBD dataset in Gama et al. [37]), while others focused on at-risk or delayed development groups, ensuring representation of both healthy and clinical samples.

3.3.2. Data Collection Methods

Data collection predominantly utilized sensor-based and imaging technologies, including inertial measurement units (IMUs) with accelerometers and gyroscopes (e.g., at 50–100 Hz in Airaksinen et al. [25,36], Franchak et al. [32,38], and Duda-Goławska et al. [40]), force plates for center of pressure (COP) analysis (e.g., 60–200 Hz in Li et al. [15] and Arias Valdivia et al. [27]), and video cameras for pose estimation (e.g., 15–60 fps in Yang et al. [26], Gama et al. [37], and Rachwani et al. [41]). Other methods involved pressure sensor mats for sitting posture (e.g., 8 × 8 grids at 10 Hz in Kim et al. [13,14] and Lee et al. [28]),

Kinect depth cameras for anthropometric measurements (in Tao et al. [34]), and custom wearables like smart jumpsuits or suits (in Airaksinen et al. [25,36]). The features extracted included kinematic variables (e.g., joint angles, velocities), pressure distributions, and time-series data from short trials (e.g., 20–30 s) or full-day recordings, often in controlled or naturalistic environments.

3.3.3. Study Settings and Research Focus

Studies were conducted in varied settings, including laboratories for controlled assessments (e.g., biomechanics labs in Li et al. [15], Arias Valdivia et al. [27], and Eken et al. [33]), home environments for naturalistic data (e.g., video recordings in Yang et al. [26], Franchak et al. [32,38], and Ledwon et al. [35]), clinical hospitals (e.g., for CP patients in Bertoncelli et al. [29] and Khaksar et al. [16]), and schools (e.g., for FHP screening in Tao et al. [34]). The primary research focus was on posture-related applications, such as classifying gross motor milestones and developmental delays in infants (e.g., Airaksinen et al. [4,20] and Yang et al. [6]), identifying sitting postures to prevent musculoskeletal issues (e.g., Kim et al. [13,14] and Lee et al. [28]), diagnosing conditions like ASD, CP, or FHP via postural sway or movement patterns (e.g., Li et al. [15], Arias Valdivia et al. [27], and Tao et al. [34]), and predicting injury risks or asymmetries (e.g., Li et al. [31] and Ledwon et al. [35]). Emphasis was placed on early detection, automated screening, and longitudinal tracking for clinical and rehabilitative purposes.

3.3.4. Machine Learning Implementation

Machine learning implementations featured a range of algorithms, with classifiers like Random Forest (e.g., achieving 94% accuracy in Yang et al. [26] and 87.75–89.39% in Khaksar et al. [16]), Support Vector Machines (SVMs; e.g., 88–97.37% accuracy in Sukhadia & Kamboj [30] and Ali & Mohamed [39]), and Convolutional Neural Networks (CNNs; e.g., 95.3–99.66% for posture classification in Kim et al. [13], Kim et al. [14], Lee et al. [28], and Airaksinen et al. [36]) being prominent. Deep learning models such as LSTM/GRU variants (e.g., 76.43% accuracy in Arias Valdivia et al. [27]) and pose estimation networks (e.g., ViTPose with 92% OKS in Gama et al. [37]) were used for temporal and video data, often with cross-validation yielding high metrics (e.g., AUC 0.865–0.98, F1-scores 0.617–0.955). Key applications included real-time posture monitoring, early disorder prediction, and injury risk assessment, with conclusions highlighting feasibility for at-home tools, clinical diagnostics, and personalized interventions, though performance varied with user familiarity and data quality. Although Table 3 reports accuracy ranges for various algorithms, direct comparison across studies is limited due to the absence of standardized evaluation metrics and heterogeneous methodologies. Differences in reported accuracies may stem from variations in sensor modalities (e.g., IMU, pressure mat, camera), data quality, preprocessing pipelines, and model architectures rather than intrinsic algorithmic superiority. Future research should adopt consistent benchmarking protocols and unified performance indicators to enable fairer cross-study comparisons of ML approaches in pediatric posture analysis.

To facilitate cross-study comparison, Table 4 summarizes the main machine learning models applied across the reviewed studies, detailing their corresponding tasks, sensing modalities, dataset characteristics, key performance indicators, and, where available, inference times. This comparative overview highlights the methodological diversity underlying reported accuracies and underscores the influence of factors such as sensor type, data volume, and computational approach on model performance.

Table 3. Main characteristics and findings of machine learning applications.

Ref.	Participants' Characteristics	Activity Registration				Aim of Prediction Related to Posture	MLe/DL Accuracy		Conclusions	Practical Application for Predicting
		Tool	Tool's Specifications	Location	Attributes/Features/Variables		Algorithm	%		
Airaksinen et al. [25]	134 infants (4–22 months); Cohort 1 (n = 97, typically developing), Cohort 2 (n = 37, developmental risk)	MAIJU wearable suit (multi-sensor)	Accelerometer and gyroscope; 52 Hz (1024 samples/s); Bluetooth to mobile device	Limbs (standard locations on each limb)	Second-by-second posture and movement detections ("MAIJU features"): supine, prone, sitting, crawling, standing, walking, etc.	To detect gross motor milestones (GMMs), quantify time spent in key postures, and track holistic motor development (BIMS score)	Support Vector Machine (SVM) for GMM detection; Linear Mixed-Effects Model (LME) for BIMS	GMM detection accuracy: 90.9–96.8% (cross-val. and external val.); BIMS vs. age correlation (Spearman's ρ): 0.93	Unsupervised at-home wearable measurements can accurately and automatically quantify infant gross motor skills, with performance comparable to expert agreement levels.	Objective, at-home assessment of infant motor development for early detection of delays, clinical support, and longitudinal tracking in healthcare and research.
Li et al. [15]	50 children (25 ASD, 25 TD); age 5–12 years; mild ASD (level 1)	Force plate (Kistler Instrument Corp., Winterthur, Switzerland)	60 Hz sampling rate; 20 s quiet standing trials; eyes open and eyes closed conditions	Biomechanics laboratory	COP linear displacements (AP, ML), total distance, sway area, sample entropy (complexity)—12 features total (6 variables × 2 conditions)	Automated identification of ASD based on postural control patterns	Naïve Bayes (best performer among 6 classifiers)	Accuracy: 0.900 Sensitivity: 0.826 Specificity: 1.000 Precision: 1.000 F1 Score: 0.898	Naïve Bayes best identified ASD postural control with high accuracy and specificity. COP complexity improved classification by ~4%.	Potential as an early diagnostic tool for ASD using postural sway biomarkers; supports computer-aided diagnosis with minimal human intervention.
Yang et al. [26]	90 infants (aged 2–6 months); 26 with developmental delays, 64 typically developing	Video camera (home recordings)	15 frames per second	Home environment	227 features extracted (106 significant after ANOVA): • Kinematic (speed, acceleration of elbows, wrists, knees, ankles) • Joint angles (shoulders, elbows, hips, knees) • Angular velocity and acceleration • Entropy of movement	Automatic classification of gross motor development (normal vs. abnormal)	Random Forest	Accuracy: 94% F1-score: 0.94 AUC: 0.98	The ViTPose model provided the best pose recognition. A Random Forest classifier using 106 significant features achieved high performance in automatically identifying infants with gross motor delays from home videos.	Enables remote, automated early screening for gross motor developmental delays in infants using simple home videos, facilitating timely intervention, especially in resource-limited areas.
Kim et al. [13]	10 children; age not specified (school-aged); no health conditions reported	Sensing cushion with pressure sensor mat	Film-type FSR sensor (8 × 8 grid); 318 × 318 mm area; 12-bit data; Bluetooth transfer	Seat cushion of a child's chair	Two-dimensional pressure distribution images (8 × 8 pixels); raw pressure values used as input features.	To classify children's sitting postures into five categories in real time	CNN (LeNet-5 modified), NB, DT, NN, MLR, SVM	CNN: 95.3% (Avg. accuracy, individual validation); NB: 87.1%; MLR: 84.5%; DT: 79.4%; NN: 92.1%; SVM: 94.2%	The CNN algorithm outperformed conventional machine learning algorithms for classifying sitting postures from pressure distribution data. Accuracy was influenced by the child's body weight.	Enables the development of a smart chair or real-time posture monitoring system to promote correct sitting habits and prevent musculoskeletal disorders in children.
Arias Valdivia et al. [27]	57 pediatric patients (7–14 years, 9.2 ± 1.8 years; 29 males, 28 females) diagnosed with hemiplegia (n = 35) or diplegia (n = 22)	AMTI force platform	Force plate; 200 Hz sampling rate; measures forces (F_x, F_y, F_z) and moments (M_x, M_y M_z)	Laboratory setting; participants stood barefoot on the platform	Center of pressure (COP) coordinates, velocity (VEL_x, VEL_y), standard deviation (STDx, STDy), area of ellipse; time series data derived from 30 s trials.	To classify type of cerebral palsy (hemiplegia vs. diplegia) based on postural control analysis during quiet standing	LSTM, GRU, BiLSTM, BiGRU, ARIMA (BiGRU performed best)	Accuracy: 76.43% (BiGRU)	BiGRU model most effectively captured temporal dependencies in postural sway for classification, outperforming traditional methods and unidirectional models; offers a non-invasive, objective diagnostic aid.	Provides a clinical decision-support tool for differentiating CP subtypes using brief, static postural control tests, aiding in early and accurate diagnosis and personalized intervention planning.
Khaksar et al. [16]	Two age groups: • ~15 years (MIT trial): 89 with CP, 30 without CP • ~3 years (iWHOT trial): 51 with CP, 20 without CP	Custom IMU (Inertial Measurement Unit)	• Sensor: MPU-9250 (Accel, Gyro, Mag.) • Microcontroller: Custom Arduino Pro Mini • Wireless: nRF24L01 (2.4 GHz RF) • Data Rate: 100 Hz • Battery: 3.7 V, 90 mAh (~3 h)	Two sensors placed: 1. Back of the hand 2. Above the wrist	Raw accelerometer and gyroscope data (time-domain) converted to frequency-domain features (Fast Fourier Transform). Feature vector (1 × 270) included amplitude, phase shift, and peak frequency of the first 5 harmonics for each sensor and axis.	To classify movement features associated with cerebral palsy (CP) from raw IMU data during a "stop sign" wrist motion task, distinguishing between children with and without CP	9 algorithms tested, including: • Random Forest • C4.5 Decision Tree • SVM, k-NN, MLP, etc.	• ~15 y.o.: 87.75% (Random Forest) • ~3 y.o.: 89.39% (C4.5 Decision Tree)	Machine learning applied to raw IMU data can successfully classify CP-related movement features without complex joint angle calculations. Decision-tree-based algorithms (Random Forest, C4.5) were most accurate. The system shows potential for accurate active range-of-motion assessment.	Provides a digital solution for classifying movement disorders like CP; could be used to monitor therapy effectiveness and for continuous at-home monitoring of hand movement patterns in children with movement disabilities.
Kim et al. [14]	26 children (14 males, 12 females); age: 6–12 years old; all physically healthy	Film-type pressure sensor (Force Sensing Resistor—FSR) array	64 (8 × 8) sensors; Sensor mat dimension: not fully specified, but sensor distance: 30 mm (w) × 30 mm (l)	Chair seat	Body pressure distribution data converted into 25 × 28 pixel heat map images.	Classification of 7 sitting postures: (a) sitting straight, (b) lean forward, (c) lean left, (d) lean right, (e) lean backward, (f) sitting at front of chair, (g) sitting crossed-legged	Convolutional Neural Network (CNN)	Accuracy: 97.5% (average from tenfold cross-validation; min: 0.970, max: 0.981). Recall and Precision for all postures > 0.9	The CNN algorithm was significantly superior (97.5% accuracy) to ANN (82.9%) and MNN (88.7%) for classifying children's sitting postures using only seat pressure sensors. This confirms the applicability of CNN-based algorithms for smart chairs to support correct posture in children.	Development of smart chairs for children to monitor sitting posture in real time, helping to prevent musculoskeletal disorders and promote the formation of correct postural habits during childhood.
Lee et al. [28]	24 healthy children (11 boys, 13 girls); age: 7–12 years (mean = 10.13, SD = 1.62); country: South Korea	Custom film-type pressure sensor mat	• 64 (8 × 8) FSRs • Size: 318 × 318 mm • Frequency: 10 Hz • Resolution: 12-bit • Data transmission: Bluetooth	Seat pan of an adjustable children's chair	2D pressure distribution maps (8 × 8 grid). Data was interpolated to 16 × 16 and normalized before being used as input.	Classification of nine sitting postures: good, leaning forward, leaning left, right foot over left, leaning right, left foot over right, sitting at front edge, slouching, crossed legs	Convolutional Neural Network (CNN)	Exp. 1 (User-specific): 99.66% Exp. 2 (All users, identifiable): 99.40% Exp. 3 (Unfamiliar user—Leave-one-out): 77.35% Good vs. Poor posture discriminator (unfamiliar user): PPV = 0.59, NPV = 0.95	A CNN applied to seat pan pressure distribution data is highly effective for classifying children's sitting postures, especially when user-specific data is available. Performance drops for unfamiliar users but remains viable for good/poor posture discrimination.	Enables the development of a non-invasive, real-time sitting posture monitoring and correction system for children in classrooms or at home, helping to prevent musculoskeletal disorders by promoting postural awareness.

Table 3. *Cont.*

Ref.	Participants' Characteristics	Activity Registration				Aim of Prediction Related to Posture	MLe/DL Accuracy		Conclusions	Practical Application for Predicting
		Tool	Tool's Specifications	Location	Attributes/Features/Variables		Algorithm	%		
Bertoncelli et al. [29]	102 teenagers with CP (60 inpatients, 42 outpatients; 60 males); mean age 16.5 ± 1.2 years (range 12–18 yrs); with cognitive impairment and severe motor disorders	Clinical and functional assessment data from medical records and standardized scales	Data collected between 2006 and 2021; variables included type of etiology, spasticity, dystonia, epilepsy, neuromuscular scoliosis, hip dysplasia, GMFCS, MACS, EDACS	Two specialized hospitals (Nice, France)	Independent variables: type of etiology (ET), sex (SE), dystonia (D), spasticity (SP), epilepsy (E), neuromuscular scoliosis (NS), hip dysplasia (H), MACS, GMFCS, EDACS. Dependent variable: type of trunk muscle tone (hypotonic, spastic, normal).	To identify factors associated with hypotonic or spastic truncal tone (TT) in adolescents with CP	Multiple Logistic Regression (TT-PredictMed model)	Average Accuracy: 82% Sensitivity: 71% Specificity: 90%	The TT-PredictMed model successfully identified specific clinical factors (e.g., hip dysplasia, etiology, motor function scores) associated with hypotonic and spastic truncal tone. The model's performance aligns with recent MLe applications in clinical diagnostics.	Enables clinicians to identify adolescents with CP at risk for specific types of postural instability (hypotonic/spastic TT), allowing for earlier, more personalized rehabilitation targeting trunk control.
Sukhadia & Kamboj [30]	43 infants (15 healthy, 28 with spastic cerebral palsy)	Custom IMU sensors (9 units)	Tri-axial accelerometer, gyroscope, and magnetometer; samples per second not specified	Nine sensors: both forearms, both upper arms, both lower legs, both upper legs, and trunk	Joint angles from limbs and trunk (3D data from 9 tri-axial sensors); stride length; leg dimension.	To detect spastic cerebral palsy (CP) based on posture and movement analysis	Support Vector Machine (SVM)	Accuracy: 88% (with 2-fold cross-validation)	An IMU-based system combined with machine learning can accurately identify infants with spastic CP by analyzing joint angles and movement parameters.	Enables early, automated detection of spastic CP using wearable sensors, facilitating timely intervention; serves as an objective tool to assist clinical diagnosis.
Li et al. [31]	Simulation study using a validated 6-year-old child occupant model (TUST IBMs 6YO-O); no human subjects	Finite Element (FE) vehicle crash simulation model; Machine Learning models	- FE Model: TUST IBMs 6YO-O (Hexahedron elements, detailed brain anatomy). - Simulation: LS-DYNA or similar explicit dynamics solver. - MLe Inputs: Collision speed (50–80 km/h), Sitting angle (90–135°). - Outputs: Head injury criteria (HIC15, 3 ms acceleration, BrIC, von Mises stress, Maxshear stress, MPS).	Virtual crash test environment (FRB—Frontal Rigid Barrier)	Input variables: impact speed, occupant sitting posture. Output (posture/injury) variables: head linear and rotational injury metrics (3ms accel., HIC15, BrIC), brain tissue stress/strain (von Mises, Maxshear, MPS).	To predict head injury risk and biomechanical response of a 6-year-old child occupant based on collision speed and sitting posture	LSTM, SVM, Random Forest (RF)	$R^2 > 0.93$ for all models (LSTM, SVM, RF) on head injury indices (e.g., 3 ms accel., HIC15) using 10-fold cross-validation	The combination of simulation and machine learning provides a reliable predictive model for child head injury. The risk and primary mechanism of injury (linear vs. rotational load) are significantly influenced by both collision speed and sitting posture.	Enables virtual safety testing and optimization of child restraint systems (CRS) in autonomous vehicle scenarios by predicting injury outcomes for various postures and crash speeds, informing safer CRS design.
Franchak et al. [32]	15 infants (6–18 months, M = 11.28 months); 7 male, 8 female; laboratory study	MetaMotionR IMUs (Mbientlab); Later: Biostamp IMUs (MC10) in leggings (home case study)	3–4 IMUs; Accelerometer and Gyroscope; 50 Hz (Lab), 62.5 Hz (Home)	Right hip, thigh, and ankle (Lab); both hips and ankles (Home, embedded in leggings)	204 features from 4 s windows: 10 summary stats (mean, SD, skew, etc.) per sensor location, signal (accel/gyro), and axis; cross-sensor/axis correlations and differences.	Classify body position into 5 categories: supine, prone, sitting, upright, held by caregiver	Random Forest	Individual Models: 97.9% (Lab), ~86% (Home) Group Models: 93.2% (Lab)	Method accurately classifies infant body position and captures individual differences in time spent in each position; feasible for contactless, full-day home assessment.	Enables unobtrusive, long-form measurement of naturalistic infant motor behavior and posture in home settings, useful for developmental studies linking motor experience to other domains (e.g., language).
Eken et al. [33]	10 children (5 with HIV encephalopathy (HIVE), 5 typically developing (TD)); all girls; aged 5–12 years; GMFCS level II (HIVE group)	Video cameras (Bonita)	50 Hz; sagittal and frontal plane recordings; DeepLabCut (pre-trained ResNet101 model) for 2D markerless pose estimation	Laboratory setting; anatomical landmarks (shoulder, elbow, wrist, hip, chin) tracked in sagittal and frontal planes	Joint angles calculated from tracked landmarks: shoulder flexion/extension, elbow flexion/extension, shoulder abduction/adduction, trunk lateral sway. Joint angle trajectories and range of motion (ROM) over the gait cycle.	To classify and quantify differences in upper body postures and movements (arm swing, trunk sway) during gait between children with and without HIVE	Statistical Parametric Mapping (SPM); Mann–Whitney U test for ROM	SPM identified significant differences in joint angle trajectories: shoulder sagittal (20–59% gait cycle), shoulder frontal (80–93%), trunk frontal (44–65%). ROM significantly larger in HIVE group for shoulder abduction ($p = 0.028$) and trunk sway ($p = 0.009$).	Markerless tracking with DeepLabCut is feasible and sensitive for quantifying pathological upper body postures and movements during gait in children with HIVE, showing increased trunk sway and altered arm swing similar to other neurological disorders.	Serves as a low-cost, accessible alternative to conventional gait analysis for assessing postural deviations in clinical settings, especially in low-to-middle-income countries; useful for monitoring disease progression and therapy effectiveness.
Tao et al. [34]	514 primary school children (aged 6–12 years); 300 with forward head posture (FHP), 214 without; from 12 public schools in Nanjing, China	Kinect depth camera (Version: Kinect 2 for Windows); InBody 370 body composition analyzer; structured questionnaire.	Kinect: accuracy 0.001 m, standing distance 2 m; InBody: standard BIA protocol; Questionnaire: Chinese PAQ-A scale	Clinical/school setting; full-body frontal and lateral views	Anthropometric: age, sex, height, body weight, BMI. Lifestyle: daily and weekly homework time. Postural: Cervical Vertebrae Angle (CVA), ear-to-shoulder distance, horizontal angle of shoulders.	To predict the risk of forward head posture (FHP) disorder	Six algorithms tested: KNN, LGBM, XGBoost, RF, LM, SVM	RF AUC = 0.865 (best performer; LM AUC = 0.640 for comparison)	The Random Forest model demonstrated superior predictive accuracy for FHP. BMI, body weight, and age were the most influential predictors, with BMI being the most important.	Provides a tool for early screening and risk assessment of FHP in school-aged children, enabling targeted interventions focusing on weight management and monitoring of sedentary behaviors (e.g., homework time).
Ledwon et al. [35]	51 healthy infants aged 6–16 weeks; full-term, Apgar score 10	Sony HDR-AS200V camera	920 × 1080 px, 60 fps	Home setting; camera 1 m above infant	Six pose-based features: HBA, NPD, TBA, TAF, BPDv, BPDh (reduced to TBA, TAF, HBA via feature selection).	Automated classification of postural asymmetry (symmetry, left, right) in infants	QDA (best performer)	92.03% accuracy; 93.26% sensitivity; AUC: 0.913	The method provides quantitative, objective assessment of postural asymmetry with high sensitivity and agreement with expert judgment.	Screening tool for infant postural asymmetry; supports early neurodevelopmental assessment and therapy monitoring without additional tools.

Table 3. *Cont.*

Ref.	Participants' Characteristics	Activity Registration: Tool	Activity Registration: Tool's Specifications	Activity Registration: Location	Activity Registration: Attributes/Features/Variables	Aim of Prediction Related to Posture	MLe/DL Accuracy: Algorithm	MLe/DL Accuracy: %	Conclusions	Practical Application for Predicting
Airaksinen et al. [36]	22 infants (mean age 6.7 months); typically developing; recruited for movement analysis	"Smart jumpsuit" with 4 wearable sensors (Movesense)	Inertial Measurement Units (IMUs): accelerometer and gyroscope; 52 Hz sampling rate; wireless via Bluetooth	Proximally on upper arms and legs (4 limbs)	Raw accelerometer and gyroscope signals (24 channels); 2.3 s windows with 50% overlap; used to classify posture and movement.	To automatically classify infant posture (prone, supine, side L/R, crawl posture) and gross body movements	Convolutional Neural Network (CNN) with iterative annotation refinement (IAR)	Posture: 94.1–99.1% UAR (depending on frame set) Movement: 71.9–82.4% UAR	The smart jumpsuit and CNN classifier achieve human-equivalent accuracy in posture and movement classification, demonstrating feasibility for automated infant movement assessment.	Enables objective, quantitative tracking of infant motor development in clinical and potentially home settings for early detection of neurodevelopmental risks.
Gama et al. [37]	• Real Infants: 2 healthy infants (1f, 1m), 8–25 weeks old, 16 videos (1440 annotated images) • Synthetic Infants: MINI-RGBD dataset, 12 synthetic infants, 1000 images each	RGB video cameras	Frame-by-frame and video input processing	In-lab/synthetic environment (supine position)	• 2D coordinates of body keypoints (e.g., eyes, shoulders, wrists, hips, knees, ankles). • Derived metrics: Object Keypoint Similarity (OKS), Average Precision/Recall (AP/AR), Neck-MidHip error, percentage of missing data/redundant detections, processing speed (fps).	To compare the performance of seven 2D human pose estimation methods for automatically estimating infant body posture from video	Seven Deep Neural Networks: AlphaPose, DeepLab-Cut/DeeperCut, Detectron2, MediaPipe/BlazePose, HRNet (BU & TD), OpenPose, ViTPose	Best Performer (ViTPose): • OKS: 0.92 (Real), 0.87 (Synth) • AP (average precision): 88.5 (Real), 73.7 (Synth) • AR (average recall): 90.9 (Real), 79.1 (Synth) • Neck-MidHip Error: ~6.0% (Real)	State-of-the-art pose estimation methods (especially ViTPose and HRNet-TD) work well on infant pose estimation without additional training. Performance varies significantly between methods in accuracy, missing/redundant detections, and speed. AlphaPose was the fastest (27 fps). DeepLabCut and MediaPipe performed poorly.	Enables automatic, markerless quantification of infant posture and movement from ordinary videos (e.g., from a smartphone). This is a key enabling technology for large-scale "in the wild" movement analysis, early screening for neurodevelopmental disorders (e.g., via GMA), and studying typical motor development.
Franchak et al. [38]	22 infants (4–14 months); 10 female, 12 male; 34 testing sessions (14 from younger group 4–7 mo, 20 from older group 11–14 mo)	MC10 Biostamp IMUs (Inertial Measurement Units)	4 IMUs; accelerometer and gyroscope; 62.5 Hz	Embedded in custom leggings on both hips (thighs) and both ankles	436 features from 4 s windows: 10 summary stats (mean, SD, skew, kurtosis, percentiles, etc.) per sensor location, signal (accel/gyro), and axis; cross-sensor/axis correlations and differences.	Classify body position into 5 categories: supine, prone, sitting, upright, held by caregiver	Random Forest	Individual Models: ~97.9% (Proximal), ~86% (Distal) Group Models: ~93.2% (Proximal)	Method accurately classifies infant body position and captures individual differences in time spent in each position. Feasible for contactless, full-day home assessment.	Enables unobtrusive, long-form measurement of naturalistic infant motor behavior and posture in home settings, useful for developmental studies linking motor experience to other domains (e.g., language).
Ali & Mohamed [39]	Infants (2–5 months post-term) from MINI-RGBD and RVI-38 datasets	RGB video cameras (Sony DSC-RX100), pose estimation software (MediaPipe, OpenPose, MeTRAbs)	25 FPS, 640 × 480 to 1920 × 1080 resolution	Clinical and home settings (supine position)	Joint angles (shoulder, elbow, hip, knee), movement velocity, acceleration, anti-gravity movements, symmetry of movement, postural variability.	Early prediction of cerebral palsy (CP) based on posture and movement patterns	SVM, NN, DT, Extra-Tree, XGBoost	MINI-RGBD: 91.67% (NN) RVI-38: 97.37% (SVM)	Pose estimation with MLe classifiers effectively distinguishes CP from typical development by quantifying movement and postural features.	Automated, non-invasive early screening for CP using widely available video recordings; suitable for home or clinical use.
Duda-Goławska et al. [40]	104 infants; longitudinal study at 4, 6, 9, 12 months; 301 visits analyzed	Xsens MTw Awinda IMUs	Accelerometer, gyroscope, magnetometer; 60 Hz (resampled from occasional 40 Hz)	Trunk and legs (optimal configuration)	1920 features from 5 groups: Statistical, Frequency, Summary, Differences, Correlations; extracted from 2 s sliding windows with 1 s overlap.	Classify infant body position into 5 classes: supine, sitting, upright, prone, hands and knees	CatBoost Classifier	F1 Scores (trunk and legs): sitting (0.942), upright (0.819), supine (0.955), prone (0.924), hands and knees (0.617).	CatBoost outperformed Random Forest. Statistical features (especially from accelerometer) were most important, followed by difference features. Sensor placement on trunk and legs was optimal.	Enables automatic, accurate monitoring of infant posture during naturalistic play; useful for assessing motor development and potentially detecting delays in lab, clinical, or home settings.
Rachwani et al. [41]	21 infants (10 girls, 11 boys) aged 6–10 months (M = 8.2 months); sitting experience ranged from 5 days to 4.75 months; all typically developing, born without complications	Video camera (home recordings via Zoom)	Resolution not specified; frame rate not specified	Home environment	Behavioral (video coded): Success in touching/grasping toy, falls, hand support, changes in base of support. Kinematic (DeepLabCut): 2D coordinates of wrist, shoulder, hip; initial trunk angle, mean trunk angle, trunk angular displacement, reach time, normalized reach path, reach velocity, straightness score.	To quantify postural control (trunk kinematics) and its relation to successful multi-directional reaching during unsupported sitting	DeepLabCut (for pose estimation) and Custom MATLAB program (for kinematic variable calculation); statistical analysis (ANCOVA) performed in SPSS (version 28)	Behavioral: Touch: 100% (both directions) Grasp: ~94% (both directions) Falls: ~3% (both directions) Kinematics: No significant differences in trunk displacement or reaching kinematics between directions or across sitting experience	All infants, including novice sitters, were successful at reaching in both directions, demonstrating functional multi-directional postural control from the onset of independent sitting. Posture became more upright with experience, but arm and trunk movements during reaching were similar regardless of sitting skill level.	Provides an objective method (video-based pose estimation) to assess functional sitting postural control via multi-directional reaching. Suggests therapeutic strategies for sitting acquisition should involve variable practice in all planes of motion from early stages, rather than a specific sequence.

Notes. MLe: machine learning; DL: deep learning; MAIJU: Motor Assessment of Infants with a Jumpsuit; ASD: autism spectrum disorder; COP: center of pressure; AP: anteroposterior; ML: mediolateral; ANOVA: analysis of variance; AUC: area under the curve; ViTPose: Vision Transformer for Pose Estimation; FSR: force sensing resister; CNN: Convolutional Neural Network; NB: Naïve Bayes Classifier; DT: Decision Tree; NN: Neural Network; MLR: Multinomial Logistic Regression; SVM: Support Vector Machine: LSTM: Long Short-Term Memory; GRU: Gated recurrent unit; BiLSTM: bidirectional long short-term memory; BiGRU: bidirectional gated recurrent unit; CP: cerebral palsy; MIT: Minimising Impairment Trial; iWHOT: Infant Wrist Hand Orthosis Trial; k-NN: k-nearest neighbors; MLP: multilayer perceptron; MNN: multi-layer neural network; PPV: positive predictive value; NPV: negative predictive value; SD: standard deviation; GMFCS: Gross Motor Function Classification System; MACS: Manual Ability Classification System; EDACS: Eating and Drinking Ability Classification System for Individuals with Cerebral Palsy; IMU: Inertial Measurement Unit; BrIC: Brain Injury Criterion; MPS: Maximum Principal Strain; KNN: K-nearest neighbor; LGBM: light gradient boosting machine; XGBoost: extreme gradient boosting; LM: linear model; HBA: Head Bend Angle; NPD: Nose Point Distance; TBA: Trunk Bend Angle; TAF: Trunk Asymmetry Factor; BPDv: Bend Point Distance horizontal; BPDh: Bend Point Distance vertical; QDA: Quadratic Discriminant Analysis; UAR: unweighted average recall; BU: Bottom-Up; TD: Top-Down; GMA: General Movement Assessment; FPS: frames per second; ANCOVA: Analysis of Covariance.

Table 4. Comparative summary of machine learning models by task, sensing modality, dataset, and performance metrics.

Task/Application	Algorithm(s)	Sensing Modality	Dataset/Sample Size	Key Metrics (Accuracy/F1/AUC/Other)	Inference/Processing Time	Reference(s)
Sitting posture classification	CNN (LeNet-5), SVM, NB	Pressure mat (8 × 8 FSR)	10 children	Acc = 95.30% (CNN); SVM = 94.20%	~20 ms per frame (real time)	Kim et al. [12]
Sitting posture classification (7 classes)	CNN	Pressure mat (8 × 8 FSR)	26 children	Acc = 97.50%; Precision > 0.90 (all classes)	~15 ms per frame	Kim et al. [13]
Sitting posture classification (9 classes)	CNN	Pressure mat (8 × 8 FSR)	24 children	Acc = 99.66% (user-specific); 77.35% (unfamiliar)	Real-time feasible	Lee et al. [28]
ASD identification	Naïve Bayes	Force plate (COP features)	50 children (25 ASD, 25 TD)	Acc = 90.00%; Sens = 82.60%; Spec = 100%; F1 = 0.90	<1 s per trial	Li et al. [14]
Gross motor development (delay detection)	Random Forest	Video (home)	90 infants	Acc = 94.00%; F1 = 0.94; AUC = 0.98	~0.5 s per segment	Yang et al. [26]
Infant posture classification (prone, supine, side, crawl)	CNN	Wearable IMUs (smart jumpsuit)	22 infants	Acc = 94.10–99.10% (UAR)	Near real time	Airaksinen et al. [35]
Gross motor milestone detection	SVM + LME	Multi-sensor wearable (MAIJU)	134 infants	Acc = 90.90–96.80%; ρ(age, BIMS) = 0.93	Real time (mobile)	Airaksinen et al. [25]
CP detection (wrist motion)	RF, C4.5 DT	IMUs (hand/wrist)	140 children (89 + 51 CP)	Acc = 87.75–89.39%	~50 ms per sample	Khaksar et al. [15]
CP subtype (hemiplegia vs diplegia)	BiGRU	Force plate (COP series)	57 children (CP)	Acc = 76.43%	N/R	Arias Valdivia et al. [27]
CP detection (pose estimation)	SVM, NN	RGB video	MINI-RGBD and RVI-38 datasets	Acc = 91.67–97.37%	~25 fps (video)	Ali & Mohamed [36]
Forward head posture risk prediction	Random Forest	Depth camera + BIA	514 school children	AUC = 0.865; Acc ≈ 86%	<1 s per case	Tao et al. [33]
Postural asymmetry detection	QDA	Video (home)	51 infants	Acc = 92.03%; Sens = 93.26%; AUC = 0.913	~0.3 s per frame	Ledwoń et al. [34]
Infant posture classification (home IMUs)	Random Forest	Wearable IMUs (leggings)	15–22 infants	Acc = 86.00–97.90% (individual models)	Real time	Franchak et al. [16,17]
Infant pose estimation benchmark	ViTPose, HRNet, AlphaPose, etc.	RGB video (2D pose estimation)	MINI-RGBD + real infant videos	OKS = 0.92; AP = 88.50; AR = 90.90 (Real)	27 fps (best model)	Gama et al. [19]
Head-injury risk prediction (sitting posture/impact)	LSTM, SVM, RF	Finite-element simulation model	Synthetic (6-year-old model)	$R^2 > 0.93$ (all indices)	N/R	Li et al. [31]

Notes. Acc: accuracy; AUC: area under the ROC curve; F1: F1-score; OKS: Object Keypoint Similarity; ρ: Spearman's correlation; Sens: sensitivity; Spec: specificity; N/R: not reported; CNN: Convolutional Neural Network; SVM: Support Vector Machine; RF: Random Forest; DT: Decision Tree; QDA: Quadratic Discriminant Analysis; IMU: inertial measurement unit; BIA: bioelectrical impedance analysis; UAR: unweighted average recall; BIMS: BABA infant motor score, BiGRU: Bidirectional Gated Recurrent Unit.

4. Discussion

Traditional pediatric posture assessment methods are subjective, intermittent, and clinic-based [7–9], while machine learning applications in pediatric contexts remain understudied despite rapid advances in adult populations. Critical gaps exist regarding which technologies and algorithms are most effective for children, whether automated systems achieve clinically meaningful accuracy for early detection, and how to enable continuous naturalistic monitoring. This gap is concerning, given that early identification of postural deviations and motor delays enables timely intervention to prevent long-term complications and improve trajectories for children with neurodevelopmental conditions [4–6]. Therefore, this systematic review examined the application of machine learning algorithms for posture-related assessments in infants and children up to 12 years old, following PRISMA guidelines [21] and established systematic review protocols [22,23]. The analysis of 22 studies revealed that sensor-based technologies combined with machine learning classifiers achieve very good to excellent accuracy (typically exceeding 85%, range: 76–99%) [42] in detecting developmental delays, classifying postures, and diagnosing neurodevelopmental conditions such as cerebral palsy and autism spectrum disorder [13–16,25–41]. The predominant technologies employed included inertial measurement units, force plates, pressure sensor mats, and video cameras, with algorithms such as Random Forest, Support Vector Machines, and Convolutional Neural Networks demonstrating superior performance across diverse applications [13–16,25–41]. These findings highlight the feasibility of automated, objective posture assessment systems for clinical diagnostics, early intervention, and continuous at-home monitoring.

4.1. Sensor Technologies and Data Collection Methods

The reviewed studies predominantly utilized three categories of sensing technologies: wearable inertial measurement units, pressure-based systems, and vision-based approaches. Wearable IMUs equipped with accelerometers and gyroscopes emerged as versatile tools, enabling longitudinal tracking of infant motor development in naturalistic home environments [25,33,37,39,41]. These devices demonstrated particular effectiveness in detecting gross motor milestones, with studies reporting detection accuracies ranging from 90.9% to 96.8% for posture classification [25]. The portability and non-invasive nature of IMU-based wearables facilitate extended monitoring periods, addressing a critical limitation of traditional clinical assessments. Validation studies have established that commercially available IMUs provide accurate measurements for infant motor assessment, with accelerometers serving as reliable tools when combined with pressure mattresses for comprehensive kinematic evaluation [43]. Recent advances in wearable sensor networks have demonstrated the feasibility of quantifying full-body movement behaviors in vulnerable populations [44,45]. Furthermore, the integration of multiple sensor locations (typically on limbs and trunk) enabled kinematic analysis of gross body movements and postural orientation that successfully classified five distinct body positions with accuracy [32,36,38]. However, it is important to note that IMU-based systems capture global movement patterns and body segment orientations rather than detailed 3D joint kinematics, which remain more accurately measured by marker-based motion capture systems [43,44].

Pressure sensor arrays and force plates offered complementary advantages for sitting posture analysis and postural control assessment. Studies utilizing film-type force-sensing resistor mats demonstrated that two-dimensional pressure distribution patterns could effectively discriminate between multiple sitting postures [13,14,28]. The conversion of raw pressure data into heat map images proved particularly suitable for Convolutional Neural Network processing, with Kim et al. [14] achieving 97.5% accuracy in classifying seven distinct sitting postures. Force plate systems provided precise quantification of center of pressure dynamics, enabling automated identification of autism spectrum disorder with 90% accuracy through analysis of postural sway characteristics [15]. Vision-based approaches, including standard video cameras and depth sensors, offered the advantage of completely contactless measurement, with pose estimation algorithms successfully extracting gross kinematic features [26,34,36,38,40,43]. Computer vision-based posture analysis systems have emerged as efficient screening tools in clinical practice, demonstrating conformity with radiographic parameters for spinal deformity detection [46].

4.2. Machine Learning Algorithms and Classification Performance

Random Forest and Support Vector Machine classifiers demonstrated robust performance across multiple studies. Random Forest achieved notable success in developmental delay classification (94% accuracy) [26] and cerebral palsy detection (87.75–89.39% accuracy) [16], likely attributable to the algorithm's ability to handle high-dimensional feature spaces and non-linear relationships. The interpretability of decision tree-based models represents an additional advantage in clinical contexts, where understanding feature importance can provide mechanistic insights into postural control deficits and support clinical decision-making [29,30]. Support Vector Machines exhibited comparable efficacy, with reported accuracies ranging from 88% to 97.37% for cerebral palsy prediction [31,40], demonstrating particular effectiveness when combined with carefully engineered frequency-domain features derived from inertial sensor data [16].

Convolutional Neural Networks emerged as the preferred architecture for image-based posture classification tasks. Multiple studies reported CNN accuracies exceeding 95% for sitting posture classification from pressure sensor arrays [13,14,28], with Lee et al. [28] achieving 99.66% accuracy for user-specific models. The superior performance of CNNs compared to traditional machine learning algorithms (Artificial Neural Networks and Multi-layer Neural Networks achieved 82.9% and 88.7%, respectively) [14] underscores the value of deep learning architectures for pattern recognition in high-dimensional sensor data. Recurrent neural network variants, including Long-Short-Term Memory (i.e., Long Short-Term Memory is an improvement over standard RNN, designed to address the difficulty of learning long-term dependencies) and Gated Recurrent Units, showed promise for temporal sequence modeling of postural sway dynamics [27]. Recent pose estimation networks, particularly vision transformer-based approaches, demonstrated state-of-the-art performance (92% Object Keypoint Similarity) for automated infant pose tracking from video [36], suggesting that transfer learning from large-scale human pose datasets can effectively address the challenges of limited pediatric training data. Comparative evaluations of deep neural network pose estimation methods (including AlphaPose, DeepLabCut, Detectron2, HRNet, MediaPipe, OpenPose, and ViTPose) on infant videos revealed that modern architectures achieve competitive performance without additional fine-tuning, with ViTPose demonstrating superior accuracy across multiple benchmarking metrics [37,46]. Analysis of end-to-end neural network architectures for infant motility assessment has demonstrated that optimization of encoder modules and data augmentation strategies can significantly enhance classifier robustness, particularly under noisy conditions in real-world scenarios [36,47].

4.3. Clinical Applications and Diagnostic Capabilities

The reviewed studies demonstrate substantial potential for machine learning-assisted early detection of neurodevelopmental disorders through posture and movement analysis. Automated identification of autism spectrum disorder from postural control patterns represents a particularly promising application, with Li et al. [15] reporting 100% specificity and 90% overall accuracy using center of pressure complexity measures. The good specificity suggests potential utility as a screening tool to rule out ASD, though the moderate sensitivity (82.6%) indicates that confirmatory assessment would remain necessary for suspected cases. Cerebral palsy detection and subtype classification emerged as another well-established application domain, with multiple studies achieving accuracy levels considered clinically meaningful for screening applications (>80% accuracy, with sensitivity and specificity adequate to reduce false negatives/positives compared to standard clinical assessment) [16,27,29,30,39]. Arias Valdivia et al. [27] successfully differentiated hemiplegia from diplegia with 76.43% accuracy using postural sway time-series data, while Khaksar et al. [16] distinguished children with cerebral palsy from typically developing controls with 89.39% accuracy in toddlers using wrist movement patterns during functional tasks.

The progression from diagnostic classification to continuous monitoring and quantitative assessment of motor development represents an emerging frontier with significant clinical implications. Airaksinen et al. [25] demonstrated that automated gross motor milestone detection could achieve performance comparable to expert clinician agreement (correlation of 0.93 between automated scores and standardized assessments), enabling longitudinal tracking of motor trajectory without repeated clinical visits. This capability addresses critical gaps in current pediatric care, where developmental surveillance often relies on infrequent clinic visits and subjective parental report. Validation studies have confirmed that quantified assessment of infant motor performance using fully automated analysis pipelines can replicate across independent cohorts from out-of-hospital recordings [25,36]. The extension of postural monitoring to preventive applications, including forward head posture screening in school populations [18] and real-time sitting posture correction systems [13,14,28], illustrates the broader applicability of these technologies beyond clinical diagnosis. The Random Forest model developed by Tao et al. [34] identified body mass index and sedentary behavior patterns as key risk factors for forward head posture (AUC = 0.865), facilitating targeted interventions for at-risk children. However, the transition from controlled research settings to clinical implementation will require validation across diverse populations, standardization of assessment protocols, and integration with existing clinical workflows to ensure that algorithmic predictions appropriately complement rather than replace clinical expertise [48,49].

4.4. Naturalistic Assessment and Home-Based Monitoring

The capability to conduct posture and movement assessment in naturalistic home environments represents a paradigm shift from traditional laboratory-based evaluations. Multiple studies successfully demonstrated that wearable sensors embedded in everyday garments (smart jumpsuits, leggings) could capture motor behavior representative of real-world activities across entire days of infant activity [25,32,36,38]. Franchak et al. [32] reported that distal sensor placement (ankles and hips) achieved 86% accuracy for body position classification during naturalistic play, while proximal placement improved accuracy to 97.9%, establishing optimal sensor configurations for at-home monitoring. The feasibility of full-day recording addresses a fundamental limitation of clinical assessments, which typically capture only brief trials that may not represent typical motor patterns.

Vision-based approaches offer the advantage of leveraging video recordings captured by parents using standard smartphones. Yang et al. [26] achieved 94% accuracy in identifying gross motor developmental delays from home videos using pose estimation combined with Random Forest classification, demonstrating that accessible technology can enable remote screening in resource-limited settings. Gama et al. [37] revealed that modern architectures (ViTPose, HRNet) achieve good pose estimation accuracy (Object Keypoint Similarity of 0.92) without requiring infant-specific training data, facilitating immediate deployment for large-scale screening initiatives. Recent analyses confirmed that pose estimation methods provide accurate tracking of gross body movements and postural configurations using low-cost imaging systems with processing speeds suitable for real-time applications, though these methods capture global movement patterns rather than detailed 3D joint kinematics [37,46]. Rachwani et al. [41] demonstrated that video-based kinematic analysis could quantify trunk stability and movement coordination in naturalistic sitting contexts, while Ledwoń et al. [35] achieved 92.03% accuracy and 93.26% sensitivity for automated postural asymmetry assessment. The convergence of wearable and vision-based approaches toward unobtrusive, longitudinal assessment creates opportunities for early identification of subtle developmental deviations that may not manifest during brief clinical examinations.

4.5. Limitations and Practical Applications

Several methodological limitations warrant consideration when interpreting the findings of this systematic review. The MINORS quality assessment revealed consistent weaknesses in prospective data collection (most studies scored 0/2 on item 3) and sample size estimation (item 8). The predominance of cross-sectional designs limits the capacity to establish predictive validity and determine whether automated assessments can reliably forecast long-term developmental outcomes, a critical requirement for clinical decision support systems. Sample sizes ranged considerably from small pilot studies with 10 participants to larger cohorts of 514 children. Many studies lacked external validation on independent datasets, raising concerns about generalizability across diverse populations, cultural contexts, and clinical settings. The performance degradation observed when machine learning models encountered unfamiliar users underscores the importance of developing person-independent classification approaches or implementing user calibration procedures. Furthermore, the heterogeneity in sensor specifications, data collection protocols, feature extraction methods, and machine learning implementations across studies precludes direct comparison of algorithmic performance and complicates the identification of optimal methodological approaches. These methodological concerns align with broader challenges in clinical AI validation, where longitudinal evaluation and standardized interpretability metrics remain essential for responsible implementation.

A critical limitation across the reviewed studies is the limited implementation and reporting of explainable artificial intelligence (XAI) methods. Among the 22 included studies, only one explicitly employed XAI techniques (Slijepcevic et al. [19]), which applied explainability methods to gait analysis in children with cerebral palsy to identify clinically relevant biomechanical patterns. Several studies using tree-based algorithms (Random Forest [16,26,27] and Support Vector Machines [30,39]) provided inherent interpretability through feature importance rankings, which identified key predictors such as body mass and sedentary behavior patterns [34]. However, most studies employing deep learning architectures—including Convolutional Neural Networks [13,14,28,36], recurrent neural networks [27], and vision transformers [37]—achieved good accuracy (>95%) without reporting explainability analyses such as GradCAM, attention visualization, or SHAP values. This lack of transparency presents a significant barrier to clinical adoption, as clinicians require an understanding of model reasoning to verify clinical relevance, identify potential biases, and maintain appropriate oversight of algorithmic recommendations [48,49]. Future research should prioritize implementing and reporting XAI methods appropriate to each algorithm type: GradCAM and saliency visualization for CNNs, attention mechanisms for temporal models, and SHAP or LIME for model-agnostic interpretability [19,48,49].

Despite these limitations, the reviewed evidence establishes clear practical applications for machine learning-assisted posture assessment in pediatric healthcare and educational settings. Smart chair systems for real-time sitting posture monitoring and wearable-based motor milestone tracking demonstrate very good to excellent accuracy for deployment in school and primary care settings. Integration with routine screening could enhance the efficiency of health programs, while continuous at-home monitoring provides objective metrics of therapy response for children with neurodevelopmental conditions. Video-based screening tools hold particular promise for expanding developmental surveillance to underserved populations. Successful clinical translation requires explainable artificial intelligence approaches that ensure transparency and interpretability of model decisions, addressing the "black box" problem that limits clinical trust and adoption. The development of clinical decision support systems incorporating explainability methods such as SHAP and LIME can enhance clinician understanding of algorithmic recommendations while maintaining diagnostic accuracy. Collaboration among researchers, clinicians, and regulatory bodies remains essential to establish validation standards, ensure data privacy protections, and develop implementation frameworks that preserve clinical judgment while leveraging the quantitative capabilities of machine learning-assisted assessment.

Future research should prioritize several critical directions. First, ML-driven adaptive interventions could transform pediatric posture management from static assessment to dynamic, personalized treatment systems. Smart chair systems [13,14,28] could evolve to provide adaptive real-time feedback based on individual response patterns, while wearable systems [25,36] could integrate with mobile applications to deliver personalized motor development activities with difficulty automatically adjusted based on progress trajectories. Such adaptive systems employing reinforcement learning and continuous monitoring could optimize intervention timing and content for maximum effectiveness while improving treatment adherence. Second, longitudinal prospective studies are needed to establish predictive validity for long-term developmental outcomes, as current cross-sectional designs cannot determine whether automated assessments provide meaningful prognostic information. Third, external validation across diverse populations is essential to establish generalizability and identify algorithmic biases. Fourth, standardization of protocols, feature extraction methods, and benchmark datasets would facilitate meaningful comparisons and accelerate methodological innovation. Finally, implementation science studies examining clinical adoption barriers, training requirements, and workflow integration are needed to translate research prototypes into deployable clinical tools.

5. Conclusions

Machine learning algorithms combined with sensor-based and vision-based technologies demonstrate substantial promise for objective, automated assessment of posture and motor development in infants and children up to 12 years old, with studies consistently achieving accuracy rates exceeding 85%, with very good to excellent accuracy (range: 76–99%) for detecting developmental delays, classifying postures, and diagnosing neurodevelopmental conditions such as autism spectrum disorder and cerebral palsy. Key technologies, including inertial measurement units, pressure sensor arrays, force plates, and video cameras, paired with algorithms such as Random Forest, Support Vector Machines, and Convolutional Neural Networks, enable diverse applications ranging from early disorder detection and real-time sitting posture monitoring to longitudinal tracking of motor milestones in naturalistic home environments.

However, critical gaps exist: only 1 of 22 studies explicitly implemented XAI methods; most employed cross-sectional designs limiting predictive validity; and external validation across diverse populations remains insufficient. Future priorities include developing ML-driven adaptive interventions for personalized treatment optimization, conducting longitudinal studies, implementing explainable AI approaches, and performing implementation science research. Successful clinical translation requires addressing the "black box" problem through explainable artificial intelligence approaches, standardizing assessment protocols across studies, validating models on independent datasets, and implementing person-independent classification strategies to ensure that algorithmic tools appropriately complement rather than replace clinical expertise while expanding developmental surveillance to underserved populations.

Author Contributions: Conceptualization, M.R.-G. and C.D.G.-C.; methodology, M.R.-G. and C.D.G.-C.; software, C.D.G.-C. and I.O.; validation, C.D.G.-C., I.O. and L.P.A.; formal analysis, M.R.-G., C.D.G.-C. and L.P.A.; investigation, M.R.-G. and I.O.; resources, C.D.G.-C. and M.R.-G.; data curation, C.D.G.-C. and L.P.A.; writing—original draft preparation, M.R.-G., C.D.G.-C. and I.O.; writing—review and editing, C.D.G.-C., I.O. and L.P.A.; visualization, C.D.G.-C. and I.O.; supervision, M.R.-G. and L.P.A.; project administration, M.R.-G. and L.P.A. All authors have read and agreed to the published version of the manuscript.

Funding: This research was partially supported by public funds received from the "Dirección General de Investigación e Innovación del Gobierno de Aragón" through the project S53_23R.

Institutional Review Board Statement: Not applicable.

Informed Consent Statement: Not applicable.

Data Availability Statement: No new data were created or analyzed in this study.

Conflicts of Interest: The authors declare no conflicts of interest.

References

1. Velázquez-Iglesias, S.; Moreno-Rodilla, V.; Curto-Diego, B.; Pérez-Robledo, F.; Llamas-Ramos, R.; Calvo-Arenillas, J.I.; Llamas-Ramos, I. Postural Analysis in Ventral and Dorsal Decubitus Babies Using Deep Learning Techniques: A Protocol Study. *J. Clin. Med.* **2025**, *14*, 3096. [CrossRef] [PubMed]
2. Franchak, J.M. Changing Opportunities for Learning in Everyday Life: Infant Body Position Over the First Year. *Infancy* **2019**, *24*, 187–209. [CrossRef]
3. Franchak, J.M.; Kretch, K.S.; Adolph, K.E. See and Be Seen: Infant–Caregiver Social Looking during Locomotor Free Play. *Dev. Sci.* **2018**, *21*, e12626. [CrossRef]
4. Maciak, M.; Koszela, K.; Beniuk, A.; Woldańska-Okońska, M. The Assessment of Postural–Motor, Coordination, and Reflex Functions in Children and Adolescents with a History of Premature Verticalization and Ontogeny Disorders in Their First Year of Life. *Children* **2024**, *11*, 1071. [CrossRef]
5. Jones, D.; Innerd, A.; Giles, E.L.; Azevedo, L.B. Association between Fundamental Motor Skills and Physical Activity in the Early Years: A Systematic Review and Meta-Analysis. *J. Sport Health Sci.* **2020**, *9*, 542–552. [CrossRef]
6. Casey, J.; Rosenblad, A.; Rodby-Bousquet, E. Postural Asymmetries, Pain, and Ability to Change Position of Children with Cerebral Palsy in Sitting and Supine: A Cross-Sectional Study. *Disabil. Rehabil.* **2022**, *44*, 2363–2371. [CrossRef]
7. Stolinski, L.; Kozinoga, M.; Czaprowski, D.; Tyrakowski, M.; Cerny, P.; Suzuki, N.; Kotwicki, T. Two-Dimensional Digital Photography for Child Body Posture Evaluation: Standardized Technique, Reliable Parameters and Normative Data for Age 7–10 Years. *Scoliosis Spinal Disord.* **2017**, *12*, 38. [CrossRef]
8. Fortin, C.; Ehrmann Feldman, D.; Cheriet, F.; Labelle, H. Clinical Methods for Quantifying Body Segment Posture: A Literature Review. *Disabil. Rehabil.* **2011**, *33*, 367–383. [CrossRef]
9. Hendry, D.; Rohl, A.L.; Rasmussen, C.L.; Zabatiero, J.; Cliff, D.P.; Smith, S.S.; Mackenzie, J.; Pattinson, C.L.; Straker, L.; Campbell, A. Objective Measurement of Posture and Movement in Young Children Using Wearable Sensors and Customised Mathematical Approaches: A Systematic Review. *Sensors* **2023**, *23*, 9661. [CrossRef] [PubMed]
10. Harsted, S.; Holsgaard-Larsen, A.; Hestbæk, L.; Boyle, E.; Lauridsen, H.H. Concurrent Validity of Lower Extremity Kinematics and Jump Characteristics Captured in Pre-School Children by a Markerless 3D Motion Capture System. *Chiropr. Man. Ther.* **2019**, *27*, 39. [CrossRef] [PubMed]
11. Zahradka, N.; Verma, K.; Behboodi, A.; Bodt, B.; Wright, H.; Lee, S.C.K. An Evaluation of Three Kinematic Methods for Gait Event Detection Compared to the Kinetic-Based 'Gold Standard. *Sensors* **2020**, *20*, 5272. [CrossRef]
12. Deng, W.; O'Brien, M.K.; Andersen, R.A.; Rai, R.; Jones, E.; Jayaraman, A. A Systematic Review of Portable Technologies for the Early Assessment of Motor Development in Infants. *npj Digit. Med.* **2025**, *8*, 63. [CrossRef] [PubMed]
13. Kim, Y.M.; Son, Y.; Kim, W.; Jin, B.; Yun, M.H. Classification of Children's Sitting Postures Using Machine Learning Algorithms. *Appl. Sci.* **2018**, *8*, 1280. [CrossRef]

14. Kim, W.; Jin, B.; Choo, S.; Nam, C.S.; Yun, M.H. Designing of Smart Chair for Monitoring of Sitting Posture Using Convolutional Neural Networks. *Data Technol. Appl.* **2019**, *53*, 142–155. [CrossRef]
15. Li, Y.; Mache, M.A.; Todd, T.A. Automated Identification of Postural Control for Children with Autism Spectrum Disorder Using a Machine Learning Approach. *J. Biomech.* **2020**, *113*, 110073. [CrossRef]
16. Khaksar, S.; Pan, H.; Borazjani, B.; Murray, I.; Agrawal, H.; Liu, W.; Elliott, C.; Imms, C.; Campbell, A.; Walmsley, C. Application of Inertial Measurement Units and Machine Learning Classification in Cerebral Palsy: Randomized Controlled Trial. *JMIR Rehabil. Assist. Technol.* **2021**, *8*, e29769. [CrossRef]
17. Mellodge, P.; Saavedra, S.; Tran Poit, L.; Pratt, K.A.; Goodworth, A.D. Quantifying States and Transitions of Emerging Postural Control for Children Not Yet Able to Sit Independently. *Sensors* **2023**, *23*, 3309. [CrossRef] [PubMed]
18. Khan, M.H.; Schneider, M.; Farid, M.S.; Grzegorzek, M. Detection of Infantile Movement Disorders in Video Data Using Deformable Part-Based Model. *Sensors* **2018**, *18*, 3202. [CrossRef]
19. Slijepcevic, D.; Zeppelzauer, M.; Unglaube, F.; Kranzl, A.; Breiteneder, C.; Horsak, B. Explainable Machine Learning in Human Gait Analysis: A Study on Children With Cerebral Palsy. *IEEE Access* **2023**, *11*, 65906–65923. [CrossRef]
20. Muralidharan, V.; Burgart, A.; Daneshjou, R.; Rose, S. Recommendations for the Use of Pediatric Data in Artificial Intelligence and Machine Learning ACCEPT-AI. *npj Digit. Med.* **2023**, *6*, 166. [CrossRef]
21. Page, M.J.; McKenzie, J.E.; Bossuyt, P.M.; Boutron, I.; Hoffmann, T.C.; Mulrow, C.D.; Shamseer, L.; Tetzlaff, J.M.; Akl, E.A.; Brennan, S.E.; et al. The PRISMA 2020 Statement: An Updated Guideline for Reporting Systematic Reviews. *BMJ* **2021**, *372*, n71. [CrossRef]
22. Rico-González, M.; Pino-Ortega, J.; Clemente, F.M.; Los Arcos, A. Guidelines for Performing Systematic Reviews in Sports Science. *Biol. Sport.* **2022**, *39*, 463–471. [CrossRef]
23. Cochrane Consumers & Communication Review Group. Data Extraction Template for Included Studies. Available online: https://opal.latrobe.edu.au/articles/journal_contribution/Data_extraction_template/6818852/1?file=12400460 (accessed on 3 November 2025).
24. Slim, K.; Nini, E.; Forestier, D.; Kwiatkowski, F.; Panis, Y.; Chipponi, J. Original Article Methodological Index for Non-Randomized Studies (Minors). *ANZ J. Surg.* **2003**, *73*, 712–716.
25. Airaksinen, M.; Gallen, A.; Taylor, E.; De Sena, S.; Palsa, T.; Haataja, L.; Vanhatalo, S. Assessing Infant Gross Motor Performance With an At-Home Wearable. *Pediatrics* **2025**, *155*, e2024068647. [CrossRef]
26. Yang, Y.-Z.; Tsai, J.-A.; Yu, Y.-L.; Ko, M.H.-J.; Chiou, H.-Y.; Pai, T.-W.; Chen, H.-J. Automatic Movement Recognition for Evaluating the Gross Motor Development of Infants. *Children* **2025**, *12*, 310. [CrossRef]
27. Arias Valdivia, J.T.; Gatica Rojas, V.; Astudillo, C.A. Deep Learning-Based Classification of Hemiplegia and Diplegia in Cerebral Palsy Using Postural Control Analysis. *Sci. Rep.* **2025**, *15*, 8811. [CrossRef]
28. Lee, Y.; Kim, Y.M.; Pyo, S.; Yun, M.H. Development of a Sitting Posture Monitoring System for Children Using Pressure Sensors: An Application of Convolutional Neural Network. *Work* **2022**, *72*, 351–366. [CrossRef]
29. Bertoncelli, C.M.; Bertoncelli, D.; Bagui, S.S.; Bagui, S.C.; Costantini, S.; Solla, F. Identifying Postural Instability in Children with Cerebral Palsy Using a Predictive Model: A Longitudinal Multicenter Study. *Diagnostics* **2023**, *13*, 2126. [CrossRef]
30. Sukhadia, N.; Kamboj, P. IMU-Based Approach to Detect Spastic Cerebral Palsy in Infants at Early Stages. *EAI Endorsed Trans. Pervasive Health Technol.* **2024**, *10*, 1–12. [CrossRef]
31. Li, H.; Wang, Y.; He, L.; Lv, W.; Cui, S.; Ruan, J.S. *Investigation of the Prediction Model and Assessment Parameters of Head Injury of Children Occupants Based on Machine Learning*; SAE International: Detroit, MI, USA, 2024; No. 2024-01-2514.
32. Franchak, J.M.; Tang, M.; Rousey, H.; Luo, C. Long-Form Recording of Infant Body Position in the Home Using Wearable Inertial Sensors. *Behav. Res. Methods* **2023**, *56*, 4982–5001. [CrossRef] [PubMed]
33. Eken, M.M.; Meyns, P.; Lamberts, R.P.; Langerak, N.G. Markerless Upper Body Movement Tracking During Gait in Children with HIV Encephalopathy: A Pilot Study. *Appl. Sci.* **2025**, *15*, 4546. [CrossRef]
34. Tao, H.; Wen, Y.; Yu, R.; Xu, Y.; Yu, F. Predictive Model Establishment for Forward-Head Posture Disorder in Primary-School-Aged Children Based on Multiple Machine Learning Algorithms. *Front. Bioeng. Biotechnol.* **2025**, *13*, 1607419. [CrossRef]
35. Ledwoń, D.; Danch-Wierzchowska, M.; Doroniewicz, I.; Kieszczyńska, K.; Affanasowicz, A.; Latos, D.; Matyja, M.; Mitas, A.W.; Myśliwiec, A. Automated Postural Asymmetry Assessment in Infants Neurodevelopmental Evaluation Using Novel Video-Based Features. *Comput. Methods Programs Biomed.* **2023**, *233*, 107455. [CrossRef]
36. Airaksinen, M.; Räsänen, O.; Ilén, E.; Häyrinen, T.; Kivi, A.; Marchi, V.; Gallen, A.; Blom, S.; Varhe, A.; Kaartinen, N.; et al. Automatic Posture and Movement Tracking of Infants with Wearable Movement Sensors. *Sci. Rep.* **2020**, *10*, 169. [CrossRef]
37. Gama, F.; Misar, M.; Navara, L.; Popescu, S.T.; Hoffmann, M. Automatic Infant 2D Pose Estimation from Videos: Comparing Seven Deep Neural Network Methods 2025. *arXiv* **2024**, arXiv:2406.17382.
38. Franchak, J.M.; Scott, V.; Luo, C. A Contactless Method for Measuring Full-Day, Naturalistic Motor Behavior Using Wearable Inertial Sensors. *Front. Psychol.* **2021**, *12*, 701343. [CrossRef] [PubMed]

39. Ali, M.M.; Mohamed, S.I. A Pose Estimation for Motion Tracking of Infants Cerebral Palsy. *Multimed. Tools Appl.* **2024**, *84*, 8261–8286. [CrossRef]
40. Duda-Goławska, J.; Rogowski, A.; Laudanska, Z.; Zygierewicz, J.; Tomalski, P. Identifying Infant Body Position from Inertial Sensors with Machine Learning Which Parameters Matter. *Sensors* **2024**, *24*, 7809. [CrossRef] [PubMed]
41. Rachwani, J.; Santamaria, V.; Ai, X.; Goldstone, H.; Kanneth, J.; Karim, N.; Schulteis, J.; Agrawal, S. Dynamic Sitting in Infants: Limits of Stability. *Gait Posture* **2023**, *102*, 210–215. [CrossRef]
42. Chicco, D.; Jurman, G. The Advantages of the Matthews Correlation Coefficient (MCC) over F1 Score and Accuracy in Binary Classification Evaluation. *BMC Genom.* **2020**, *21*, 6. [CrossRef]
43. Rihar, A.; Mihelj, M.; Pašič, J.; Kolar, J.; Munih, M. Infant Trunk Posture and Arm Movement Assessment Using Pressure Mattress, Inertial and Magnetic Measurement Units (IMUs). *J. Neuroeng. Rehabil.* **2014**, *11*, 133. [CrossRef] [PubMed]
44. Walmsley, C.P.; Williams, S.A.; Grisbrook, T.; Elliott, C.; Imms, C.; Campbell, A. Measurement of Upper Limb Range of Motion Using Wearable Sensors: A Systematic Review. *Sports Med.—Open* **2018**, *4*, 53. [CrossRef] [PubMed]
45. Jeong, H.; Rogers, J.A.; Xu, S. Continuous On-Body Sensing for the COVID-19 Pandemic: Gaps and Opportunities. *Sci. Adv.* **2020**, *6*, eabd4794. [CrossRef]
46. Roggio, F.; Trovato, B.; Sortino, M.; Musumeci, G. A Comprehensive Analysis of the Machine Learning Pose Estimation Models Used in Human Movement and Posture Analyses: A Narrative Review. *Heliyon* **2024**, *10*, e39977. [CrossRef]
47. Airaksinen, M.; Vanhatalo, S.; Räsänen, O. Comparison of End-to-End Neural Network Architectures and Data Augmentation Methods for Automatic Infant Motility Assessment Using Wearable Sensors. *Sensors* **2023**, *23*, 3773. [CrossRef]
48. Sadeghi, Z.; Alizadehsani, R.; Cifci, M.A.; Kausar, S.; Rehman, R.; Mahanta, P.; Bora, P.K.; Almasri, A.; Alkhawaldeh, R.S.; Hussain, S.; et al. A Review of Explainable Artificial Intelligence in Healthcare. *Comput. Electr. Eng.* **2024**, *118*, 109370. [CrossRef]
49. Antoniadi, A.M.; Du, Y.; Guendouz, Y.; Wei, L.; Mazo, C.; Becker, B.A.; Mooney, C. Current Challenges and Future Opportunities for XAI in Machine Learning-Based Clinical Decision Support Systems: A Systematic Review. *Appl. Sci.* **2021**, *11*, 5088. [CrossRef]

Disclaimer/Publisher's Note: The statements, opinions and data contained in all publications are solely those of the individual author(s) and contributor(s) and not of MDPI and/or the editor(s). MDPI and/or the editor(s) disclaim responsibility for any injury to people or property resulting from any ideas, methods, instructions or products referred to in the content.

bioengineering

MDPI

Article

Effects of a Personalized Augmented Reality Exercise Program Based on Basic Fitness on Key Components of Physical Fitness in Healthy Adults: A Randomized Controlled Trial

Jaewon Lee and Jaeho Yu *

Department of Physical Therapy, Sunmoon University, Asan 31460, Republic of Korea; zzxvv99@gmail.com
* Correspondence: naresa@sunmoon.ac.kr

Abstract

Background: Augmented reality (AR)-based exercise offers a low-cost option for home training, but the lack of supervision increases the risk of incorrect performance. Therefore, AR systems must demonstrate accuracy and effectiveness comparable to therapist-led training. To address this need, this study examined whether an AR-supported multi-component exercise program improves six key components of physical fitness. Methods: Twenty-eight healthy adults from South Korea were recruited and randomly assigned to either the AR group or the physical therapist (PT) group. Participants were assessed for six components of physical fitness: muscular strength, muscular endurance, balance, cardiorespiratory endurance, flexibility, and coordination. Each group performed the exercise program three times per week for four weeks, with each session lasting at least 30 min. Statistical analyses were conducted using paired *t*-tests to evaluate pre- and post-intervention effects within each group and independent *t*-tests to compare differences between the two groups. The level of statistical significance was set at $p < 0.05$. Results: Exploratory analyses showed that both groups demonstrated post-intervention improvements in muscular endurance, balance, flexibility, and coordination, while cardiorespiratory endurance and grip strength showed no notable changes. Between-group comparisons revealed significant differences only in right-hand relative grip strength at both baseline and post-intervention, with no other domains differing between groups. Conclusions: First, the AR-based intervention significantly improved muscular endurance, balance, flexibility, and coordination. Second, there were no significant differences between the AR-based and therapist-led interventions. Consequently, AR-based exercise programs may enhance specific components of physical fitness and could be effectively utilized to complement and extend therapist-led training sessions.

Keywords: augmented reality; physical fitness; physical factors; multicomponent training program

1. Introduction

Exercise and rehabilitation are sometimes restricted by environmental factors such as weather, location, and transportation. Advances in science and technology have minimized the impact of these limitations, and particularly, virtual and augmented reality (AR) technologies have emerged as novel approaches for promoting health [1]. AR-based rehabilitation environments go beyond conventional exercise methods by providing users with new and engaging experiences, which may increase their motivation and adherence to physical training. During this process, physical therapists can analyze exercise data to objectively evaluate users' physical outcomes and adjust exercise plans accordingly [2]. The expansion of AR-based exercise sessions also enables individuals to engage in physical training at home at a relatively low cost [3]. However, home-based exercises without professional supervision make immediate correction difficult and may increase the risk of performing movements incorrectly [4]. To overcome these challenges, ongoing research has focused on encouraging user participation while maintaining accuracy and efficiency comparable to traditional supervised exercise systems [5].

According to previous studies, various forms of feedback—such as auditory, visual, textual, and video-based feedback—have been shown to enhance user motivation and facilitate not only physical activity but also cognitive learning [5]. In particular, auditory feedback demonstrated superior effectiveness in guiding users toward accurate movements [5]. Moreover, an analysis of nine AR systems incorporating different types of feedback for shoulder rehabilitation revealed clear advantages over conventional rehabilitation methods in terms of usability, enjoyment, motivation, and exercise performance outcomes [6]. However, this does not imply that traditional exercise or rehabilitation methods are ineffective. Exercise programs that include strength and flexibility training performed at appropriate intensities and frequencies are effective in delaying neuromuscular capacity decline, preventing or reducing sarcopenia, and minimizing age-related changes in body composition [7]. Therefore, combining augmented reality technology with traditional exercise methods may yield more effective outcomes.

Maintaining good health is essential, and as aging progresses, physical fitness and self-efficacy play a crucial role in improving quality of life and overall well-being. Therefore, preventive exercise interventions should be initiated before the onset of disease to promote health [8]. Physical fitness comprises multiple components: (1) muscular components such as strength and endurance, (2) motor components such as balance, (3) cardiorespiratory components, (4) metabolic components, and (5) morphological components such as height and subcutaneous fat [9]. These elements are closely associated with an individual's health status and level of physical activity. Accordingly, exercise programs should be systematically designed to promote these components of physical fitness for the maintenance and enhancement of overall health.

Multi-component exercise is a comprehensive intervention method that combines several elements of physical fitness, such as muscular strength, flexibility, and cardiorespiratory endurance. Compared to conventional single-component exercise programs, it has been shown to exert more positive effects on fitness outcomes and plays a key role in improving quality of life and preventing falls [7,10]. In a study involving older adults residing in long-term care facilities, progressive multi-component exercise was proven effective in preventing declines in cardiopulmonary health [11]. Advances in science and technology have further enhanced the implementation of such exercises by providing effective tools to facilitate them. Exercise programs in the form of video games—commonly referred to as exergames—that include strength and balance training have been found to provide greater motivation to users and contribute positively to both physical and cognitive function [12].

Previous studies have demonstrated that multi-component exercise can produce various physiological benefits that contribute to improved physical fitness. In particular, continuous research is needed to verify the effectiveness of AR-based rehabilitation environments in order to expand and enhance therapeutic sessions. Therefore, the purpose of this study was to investigate the effects of an AR-based multi-component exercise program on six components of physical fitness—muscular strength, muscular endurance, balance, cardiorespiratory endurance, flexibility, and coordination—and to test the hypothesis that this intervention would be effective in improving these components.

2. Materials and Methods

2.1. Study Design

Participants were recruited through public advertisements and were screened for eligibility by the principal investigator A according to predefined inclusion criteria. Randomization was performed by an independent assistant using a simple randomization procedure based on the RANDBETWEEN function in Microsoft Excel, which generated random allocation codes assigning participants to either the augmented reality group (ARG) or the physical therapist-led group (PTG). Allocation was concealed from investigator A; group assignments were disclosed only after each participant arrived for the first visit, at which point the assistant directed the participant to the assigned intervention room. The study took place at SUNMOON University in Asan-Si, South Korea, within the Health and Medical Sciences Building. The ARG intervention was delivered in Room 104 and the PTG intervention in Room 114, each administered by separate interventionists (C and D). Because recruitment materials described the study only as a physical fitness enhancement program and interventions were conducted in separate rooms by different interventionists, participants were unaware of their group assignment and had minimal opportunity to encounter members of the other group. All outcomes were assessed by assessor B, who remained blinded to group allocation throughout the study. This study was approved by the Institutional Review Board of Sunmoon University (IRB No. SM-202206-030-2) and was prospectively registered in the Clinical Research Information Service (CRIS; KCT0008368).

2.2. Participants

A total of 28 healthy adults were enrolled in this exploratory randomized controlled trial, which was conducted from October 2022 to February 2023. Although a formal confirmatory power calculation was not required for this exploratory design, we conducted a preliminary feasibility check using G*Power (version 3.1.9.7) [13] to ensure that the planned sample size was reasonable for detecting meaningful changes in key outcomes. Based on this assessment and practical considerations for feasibility, a target sample size of approximately 24–28 participants were deemed appropriate, and 28 participants were ultimately recruited. After eligibility screening based on predefined inclusion and exclusion criteria, all 28 individuals were randomized, and all completed the 4-week intervention without loss to follow-up or protocol deviations.

Participants were eligible if they met the following criteria: (1) no history of musculoskeletal surgery, trauma, or neurological disorders within the past six months; (2) no cardiovascular or pulmonary diseases; (3) absence of pain that could restrict exercise; (4) no psychiatric or cognitive disorders within the past two years; and (5) no vestibular dysfunction. Participants who experienced discomfort during the intervention were allowed to rest; those unable to continue due to COVID-19 infection were excluded. All participants provided written informed consent before enrollment.

Baseline height was measured using an automatic BMI measuring instrument (BSM 370, Republic of Korea, 2011), and weight was measured using a body composition analyzer (InBody 570, Biospace, Republic of Korea, 2013). Physical characteristics are summarized in Table 1.

Table 1. General characteristics of participants.

	ARG (n = 14)	PTG (n = 14)
Age (years)	22.29 ± 1.68	22.86 ± 3.01
Height (cm)	169.33 ± 8.18	162.66 ± 5.96
Weight (kg)	70.61 ± 16.08	59.33 ± 10.76

Values indicate mean ± standard deviation, ARG: augmented reality exercise group, PTG: physical therapist exercise group.

2.3. Experiment Procedures

The study procedure is illustrated in [Figure 1]. Participants underwent baseline assessments followed by randomization to either the ARG or PTG. All participants were evaluated twice—before and after the intervention—for six physical fitness components: muscular strength, muscular endurance, balance, cardiorespiratory endurance, flexibility, and coordination.

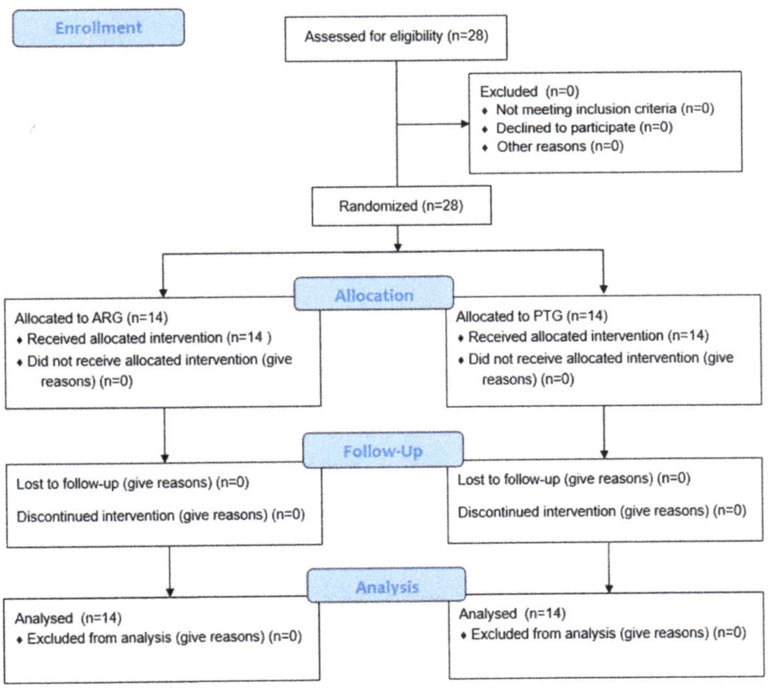

Figure 1. Consort Flowchart.

The exercise program was conducted three times per week for four weeks, totaling 12 sessions, with each session lasting 30 min. Stretching exercises were performed for three sets, maintaining each position for 15 s, while each strengthening exercise was performed for three sets of 12 repetitions. The intensity of the exercise tools used during the sessions was progressively increased each week to correspond to a Borg Rating of Perceived Exertion of 13–15.

2.4. Measurement Tools and Methods

All participants were measured for height and weight once before the intervention. Assessments of muscular strength, muscular endurance, balance, cardiorespiratory endurance, flexibility, and coordination were conducted twice—once before and once after the intervention. To minimize measurement error, all assessments were performed by the same researcher using standardized procedures and testing environments.

2.4.1. Measuring Strength

Muscular strength was assessed using a handgrip dynamometer (Jamar®, Preston, CT, USA). Participants were seated comfortably in a chair with the test arm slightly abducted at approximately 15° from the trunk and the elbow fully extended, ensuring that the arm did not touch the body. Following the researcher's verbal instruction, participants were asked to maintain the position and exert maximal grip force for 3 s. The researcher then recorded the value. Handgrip strength was measured twice alternately for both the right and left hands, and the highest value was recorded to the nearest 0.1 kg. Relative grip strength was calculated by dividing the measured grip strength by the participant's weight [14].

2.4.2. Measuring Muscular Endurance

Muscular endurance was assessed using the 30-s chair stand test (30CST), performed on a chair with a backrest but no armrests. Participants sat in the middle of the chair with their back straight, feet flat on the floor, and arms crossed at the wrists and placed across the chest. Upon the researcher's signal, participants stood up fully and then returned to a seated position. To ensure proper form, each participant practiced one or two times before the test. They were then instructed to perform as many full stands as possible within 30 s. The total number of completed sit-to-stand cycles within the 30-s period was recorded [15].

2.4.3. Measuring Balance

Balance was evaluated using the Timed Up and Go (TUG) test. One cone and one chair with a backrest but no armrests were prepared, positioned to face each other. The cone was placed exactly 3 m away from the front edge of the chair, ensuring a straight line between the two. Participants sat upright in the middle of the chair with their back straight, feet flat on the floor, and hands resting on their thighs. One foot was placed slightly ahead of the other, with the trunk leaning slightly forward. At the researcher's signal, participants stood up, walked as quickly as possible around the cone, and returned to sit on the chair. After one practice trial, participants performed the test twice, and the fastest time was recorded to the nearest 0.1 s [16].

2.4.4. Measuring Cardiopulmonary Endurance

Cardiorespiratory endurance was assessed using the 2-Minute Step Test (2MST). Upon the examiner's signal, participants began stepping in place, starting with the right foot. One full step was counted when both feet alternately touched the ground. The target knee height for each participant was individually determined by measuring the distance from the midpoint of the patella to the anterior superior iliac spine using a tape measure; participants were instructed to raise their knees to at least the midpoint level of the thigh. If the knee height was not maintained, participants were reminded to correct their posture, but the step was not counted while the timer continued. Participants were instructed to perform as many steps as possible within the 2-min period [17].

2.4.5. Measuring Flexibility

Flexibility was assessed using the Sit-and-Reach Test (SRT). Participants removed their shoes and sat upright with their knees fully extended and the soles of their feet placed flat against the vertical surface of the measurement box. The distance between the feet was kept within 5 cm. With both arms extended forward, participants reached as far as possible while keeping their knees straight and avoiding any bouncing or jerking motion.

The fingertips of both hands were required to remain aligned throughout the movement. The test was performed twice, and the better score was recorded to the nearest 0.1 cm [18].

2.4.6. Measuring Coordination

Coordination was evaluated using the Figure 8 walking test (F8W). A rectangular course measuring 3.6 m in length and 1.6 m in width was marked on the laboratory floor, with cones placed at both rear corners and a chair positioned 2.4 m from each cone. Participants sat on the front center edge of the chair and began the test upon the examiner's signal. They stood up, walked around the cone positioned diagonally to the right, returned to sit down, then immediately stood up again and walked around the cone diagonally to the left before sitting back down. The total time required to complete the task was recorded to the nearest 0.1 s. No practice trials were provided, and the use of any assistive devices was not permitted [19].

2.5. Intervention Method

Participants in the ARG performed the exercises using an AR-device [Figure 2]. A markerless RGB camera was used to project each participant's body onto the screen and identify anatomical landmarks such as joint positions. The physical fitness assessment was delivered in a fixed sequence ranging from muscle strength to coordination, accompanied by guided videos and on-screen text to ensure that even first-time users could easily follow the instructions. More than 500 stored exercises were categorized by difficulty level (high, medium, low) and by target joints or functional components. Using cut-off values determined according to the age-specific grading standards of Korea's National Fitness Award program, the system automatically assigned the appropriate difficulty level and generated an exercise program tailored to each participant's assessment results [20]. The therapist then established training goals across six components, and a unified protocol was developed for all participants as presented in Table 2. The exercise program and prescribed repetitions generated by the ARG system were applied identically to the PTG, while training intensity was individually adjusted according to assessment outcomes and the therapist's clinical judgment. During the exercises, the system provided audiovisual feedback and recorded accuracy scores in real time.

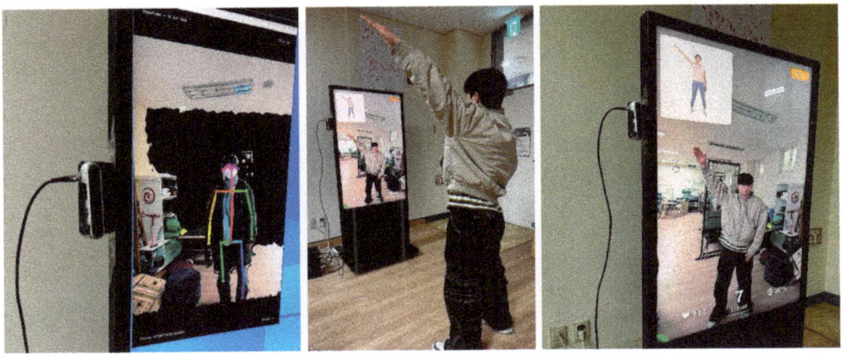

Figure 2. Augmented reality-based intervention method.

Participants in the PTG performed the exercises under the supervision of a therapist using traditional one-on-one methods. Before the intervention, the researcher explained the entire procedure and exercise program to all participants and recommended wearing comfortable clothing to ensure accurate and safe performance. The experimental environments for the two groups were assigned separately to prevent overlap, and the researcher continuously monitored both settings to maintain consistent and safe conditions.

Table 2. Exercise program.

Name of the Exercise	Explanation of Exercise
Standing shoulder extension	The subjects stand up straight with their shoulders back and their legs spread shoulder width apart. They place their hands behind their back with their palms facing backward and interlace their fingers. Then, they slowly lift their arms up and hold the position (15 s holding × 3 sets).
Standing dumbbell bicep curls with one arm bent	Stand up straight with your back straight and your feet shoulder-width apart. Keep your arms close to your sides with your palms facing forward and grab the dumbbells. Bend your elbows and lift the dumbbells up. Return to the starting position and repeat (12 times × 3 sets).
Standing forward bend	Stand up straight with your back straight and your arms at your sides. Slowly bend your head and torso while keeping your arms close to your sides until your fingertips touch the ground. Then return to the starting position (15 s holding × 3 sets).
One-legged Medicine Ball Diagonal Lift	Stand with a straight back and grab a medicine ball with both hands. Balance on one leg and bend the hip and knee joint on the supporting side, lowering the medicine ball in a diagonal direction. Straighten the supporting leg and simultaneously lift the medicine ball upwards towards the opposite diagonal direction, bending the hip and knee joint on the same side (12 times × 3 sets × each side)
Sandbag Ankle Steps	Put sandbags on your ankles and stand comfortably. Then, lift your knee up to the middle of your thigh and step up and down with each foot, as if climbing stairs (2 min × 3 sets).
Stand on one foot on a step box and lift your knee up	Stand on a step box with one leg and bend the hip and knee joints of the other leg while stepping up onto the box. Then return to the starting position and repeat with the other leg (12 times × 3 sets × each side)
Standing Calf Stretch with knee extension	Stand with your feet placed apart. Maintain the heel of the rear foot not to fall, then slowly bend the back knee and maintain the feeling of pulling the calf muscle of the rear foot (15 s holding × 3 sets)

2.6. Data Analysis

Descriptive statistics were used to calculate the general characteristics of the participants, and the mean and standard deviation were computed for all variables. All statistical analyses were performed using IBM SPSS Statistics version 26.0 (IBM Corp., Armonk, NY, USA). To evaluate the effects of the exercise program before and after the intervention within each group, paired *t*-tests were conducted, and independent *t*-tests were used to compare differences between the two groups. The level of statistical significance was set at $\alpha = 0.05$.

3. Results

Normality assumptions were confirmed prior to analysis using the Shapiro–Wilk test. At baseline, a significant difference was observed between the two groups in right-hand relative grip strength ($p < 0.05$), while no significant differences were found in any other variables ($p > 0.05$) [Figure 3, Table 3].

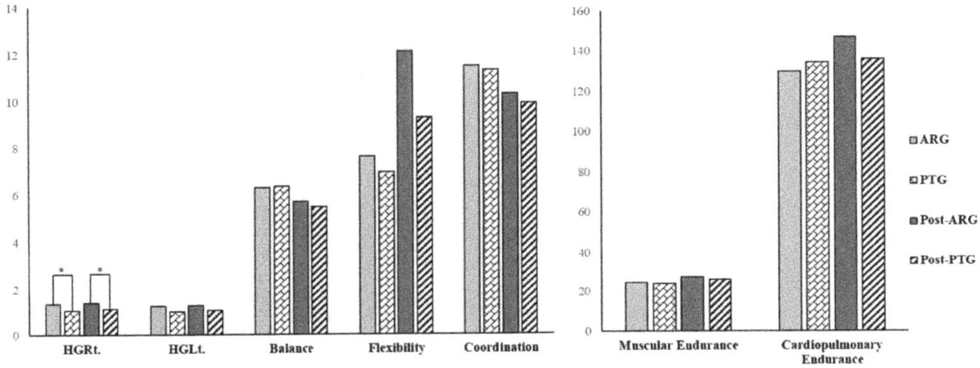

Figure 3. Comparison between groups after intervention; * $p < 0.05$, HGRt: hand grip right side, HGLt: hand grip left side, ARG: augmented reality exercise group, PTG: physical therapy exercise group.

Table 3. Descriptive Statistics and Between-Group Comparisons at Pre- and Post-Intervention.

		Pre			Post		
		ARG	PTG	t	ARG	PTG	t
Strength							
Hand grip	**Rt**	1.34 ± 0.42	1.02 ± 0.21	2.556 *	1.38 ± 0.41	1.10 ± 0.25	2.198 *
	Lt	1.24 ± 0.45	0.97 ± 0.26	1.946	1.25 ± 0.40	1.02 ± 0.25	1.825
Muscular Endurance		24.43 ± 6.87	23.71 ± 4.76	0.320	27.29 ± 7.10	26.00 ± 5.35	0.541
Balance		6.30 ± 0.95	6.35 ± 1.12	−0.106	5.72 ± 0.82	5.49 ± 0.68	0.826
Cardiopulmonary Endurance		129.79 ± 29.10	134.29 ± 18.17	−0.491	146.93 ± 43.59	136.00 ± 24.06	0.821
Flexibility		7.63 ± 11.15	6.98 ± 9.34	0.167	12.11 ± 8.77	9.30 ± 9.39	0.816
Coordination		11.49 ± 2.11	11.33 ± 1.88	0.206	10.31 ± 1.49	9.91 ± 1.12	0.802

* $p < 0.05$, mean ± standard deviation, Rt: right, Lt: left, ARG: augmented reality exercise group, PTG: physical therapy exercise group.

After the intervention, no significant differences were observed in relative grip strength (muscular strength) for either hand in both groups ($p > 0.05$). Muscular endurance significantly improved in both groups ($p < 0.01$). Balance showed significant improvement in both the ARG ($p < 0.05$) and the PTG ($p < 0.01$). No significant changes were found in cardiorespiratory endurance in either group ($p > 0.05$). Flexibility significantly improved in both groups ($p < 0.01$). Coordination also improved significantly in the ARG ($p < 0.001$) and in the PTG ($p < 0.01$) [Figure 4, Table 4].

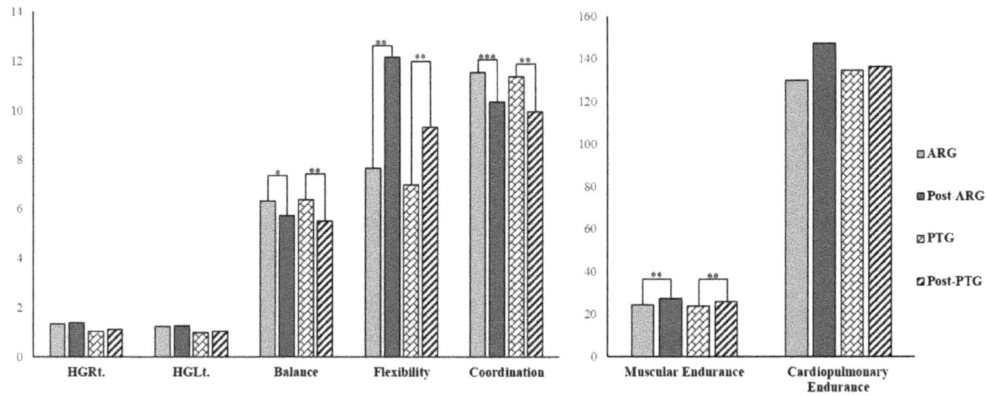

Figure 4. Comparison within a group after intervention.; * $p < 0.05$, ** $p < 0.01$, *** $p < 0.001$, HGRt: hand grip right side, HGLt: hand grip left side, ARG: augmented reality exercise group, PTG: physical therapy exercise group.

Table 4. Comparison within a group after intervention.

		ARG			PTG		
		Pre	Post	t	Pre	Post	t
Strength							
Hand grip	**Rt**	1.34 ± 0.42	1.38 ± 0.41	−0.843	1.02 ± 0.21	1.10 ± 0.25	−1.693
	Lt	1.24 ± 0.45	1.25 ± 0.40	−0.144	0.97 ± 0.26	1.02 ± 0.25	−1.105
Muscular Endurance		24.43 ± 6.87	27.29 ± 7.10	−3.899 **	23.71 ± 4.76	26.00 ± 5.35	−3.829 **
Balance		6.30 ± 0.95	5.72 ± 0.82	2.891 *	6.35 ± 1.12	5.49 ± 0.68	3.770 **
Cardiopulmonary Endurance		129.79 ± 29.10	146.93 ± 43.59	−1.864	134.29 ± 18.17	136.00 ± 24.06	−0.265
Flexibility		7.63 ± 11.15	12.11 ± 8.77	−3.898 **	6.98 ± 9.34	9.30 ± 9.39	−3.872 **
Coordination		11.49 ± 2.11	10.31 ± 1.49	4.474 ***	11.33 ± 1.88	9.91 ± 1.12	3.887 **

* $p < 0.05$, ** $p < 0.01$, *** $p < 0.001$, mean ± standard deviation, Rt: right, Lt: left, ARG: augmented reality exercise group, PTG: physical therapy exercise group.

After the intervention, a significant difference between groups was found in right-hand relative grip strength ($p < 0.05$), while no other variables showed significant between-group differences ($p > 0.05$) [Figure 3, Table 3].

4. Discussion

Many studies have reported that low muscle function is associated with higher mortality and morbidity rates [21]. Handgrip strength is widely used as a surrogate measure of overall muscle strength because it is simple, cost-effective, and easy to administer [22]. It has shown a strong association with the incidence and mortality of various diseases—including cardiovascular disease, respiratory disease, chronic obstructive pulmonary disease and cancer—and weaker grip strength is linked to poorer health outcomes, making it a reliable prognostic indicator [21,23]. Moreover, the predictive ability of grip strength for mortality remains consistent regardless of whether it is expressed in absolute or relative terms [24]. In the present study, although grip strength increased in both hands after the intervention in both groups, the improvement was not statistically significant. This finding may be explained by the similarity between the evaluated task and the type of exercise performed; greater effects are typically observed when the exercise closely matches the test variable. It has been recommended that to increase grip strength specifically, individuals should train at approximately 75% of one-repetition maximum for at least nine weeks [25]. However, studies reporting large effects often adopt a multi-component exercise approach, suggesting that integrating balance, flexibility, and endurance training may yield additional benefits for grip strength improvement [26]. Therefore, continued participation in this program or the addition of grip strength–specific exercises after reassessment may further enhance results. Additionally, this study did not stratify participants based on handedness or sex, as the primary goal was to verify the effectiveness of the AR system and the exercise protocol. Since most participants were right-handed, the authors speculate that this may have contributed to the pre- and post-intervention differences observed between groups.

Exercises included in this intervention, such as ankle sandbag kicks and step training, were sufficient to improve muscular endurance. A previous 5-week AR-based exergame intervention for community-dwelling older adults significantly enhanced lower-limb strength and functional balance performance, consistent with the findings of the present study [27]. Because lower-limb function is closely associated with balance, the 30CST is often used not only to evaluate lower-limb strength and muscular endurance but also to assess balance ability and predict fall risk [28,29]. Furthermore, when an augmented reality–based Otago exercise program was applied to older adults at high risk of falls residing in long-term care facilities, the AR group showed greater median improvements in both the Berg Balance Scale and 30CST scores compared with the conventional walking group [30].

As mentioned earlier, lower-limb function is closely related to balance. Groups with reduced lower-limb function, such as those experiencing sarcopenia and associated muscle loss, have demonstrated poorer performance on the TUG test [31,32]. A recent meta-analysis reported that AR–based exercise interventions effectively improved balance and gait ability, showing particularly strong effects in elderly populations with neurological disorders such as stroke and Parkinson's disease [33]. This may be attributed to the fact that older adults typically rely heavily on visual information; thus, AR-based exercise enhances proprioceptive feedback in the lower limbs, leading to improved balance control. These findings are consistent with the results of the present study [34,35].

Modern individuals who spend prolonged periods sitting are particularly susceptible to excessive strain in the hamstring muscles [36]. The SRT has been shown to provide a valid and reliable measure for assessing hamstring flexibility [18], and the calf muscles are also known to influence SRT performance [37]. In this study, static stretching interventions targeting the hamstring and calf muscles were effective, consistent with findings from previous research [38]. Furthermore, a study involving 68 healthy adults demonstrated that remote rehabilitation performed four times per week for eight weeks produced significantly greater improvements in flexibility compared with a conventional home exercise group [39]. Similarly, the results of the present study align with previous findings showing that both a 4-week AR-based proprioceptive exercise program and a 12-week AR-based sarcopenia prevention program significantly improved hamstring flexibility [34,40].

The 2MST is one of the commonly used assessments for cardiorespiratory endurance, similar to the 6-Minute Walk Test. It is more practical in limited physical spaces and allows rapid evaluation of aerobic capacity [41,42]. However, several studies have demonstrated that the 2MST is also correlated with lower-limb strength and muscular endurance, as well as with dynamic balance measured by the TUG test, indicating its close association with functional mobility [42–44]. Previous research has shown that AR-based exercise interventions improved cardiorespiratory endurance in older adults with reduced aerobic capacity [45]. In contrast, the present study found no statistically significant improvement in cardiorespiratory endurance, despite significant changes in other variables. This may be due to insufficient exercise duration. According to the American College of Sports Medicine (ACSM) guidelines, only sessions lasting at least 10 min can be accumulated toward total exercise time, although bouts shorter than 10 min may still provide beneficial adaptations when performed as part of high-intensity aerobic training [46]. Therefore, it may be necessary to increase the proportion and duration of aerobic components in future exercise protocols.

The F8W requires participants to perform both straight and curved paths, including clockwise and counterclockwise turns. Thus, individuals must integrate multiple sensory inputs and plan goal-directed movements to navigate complex spatial configurations. This test also demands coordination of lower-limb function and cognitive processing to smoothly transition between movement patterns [47]. AR interfaces provide significant visual advantages by facilitating the relationship between visual attention and the brain's visual information processing [48]. In this study, projecting the user's image onto the screen offered visual feedback necessary for interaction, while guided exercise videos supported initial motor learning for beginners [6]. Previous studies have shown that similar AR interfaces effectively alleviated phantom limb pain in amputees [49,50]. According to social cognitive theory, which emphasizes individual learning through continuous interaction with the environment, behavior is fundamentally learned and influenced by two key cognitive factors—self-efficacy and outcome expectancy—that determine behavioral adherence [51]. Self-efficacy plays a crucial role in promoting participation in physical activity, and AR-based environments provide strong intrinsic motivation by enhancing users' sense of agency and engagement [52].

This study has several limitations. First, participants' daily activities outside the experimental environment could not be fully controlled, potentially influencing the outcomes. Second, because the exercise program was individualized based on baseline assessments, some variables may not have shown significant changes. Third, as the participants were healthy adults in their twenties, the generalizability of the findings to other age groups or clinical populations is limited. Although previous studies have reported that AR-based exercise interventions can improve cognitive and physical functions across diverse age groups and clinical conditions—such as stroke, Parkinson's disease, and dementia—future studies should include broader age ranges and clinical samples and incorporate long-term follow-up assessments to evaluate sustained effects. Finally, despite careful planning, the relatively small sample size (14 participants per group) means that the study was underpowered to rule out modest between-group differences, particularly in the presence of baseline imbalance. Therefore, nonsignificant between-group results should be interpreted with caution and validated in larger, adequately powered trials.

5. Conclusions

This study aimed to investigate the effects of an AR-based multi-component exercise program on six physical fitness factors—muscular strength, muscular endurance, balance, cardiorespiratory endurance, flexibility, and coordination. The findings were as follows: First, the AR-based intervention significantly improved muscular endurance, balance, flexibility, and coordination. Second, the magnitude of improvement observed in the AR group was generally comparable to that of the therapist-led group, although no statistically significant between-group differences were detected. Given the pilot nature and limited statistical power of this study, these non-significant differences should not be interpreted as evidence of equivalence or noninferiority between the two interventions. In conclusion, AR-based exercise programs show promise for enhancing specific components of physical fitness and may serve as a supplementary modality to therapist-led sessions; however, larger and adequately powered trials are required to confirm comparative effectiveness.

Author Contributions: Conceptualization, J.Y.; methodology, J.L.; formal analysis, J.L.; data curation, J.L.; writing—original draft preparation, J.L.; writing—review and editing, J.L.; visualization, J.L.; supervision, J.Y. All authors have read and agreed to the published version of the manuscript.

Funding: This research received no external funding.

Institutional Review Board Statement: This study was approved by the Institutional Review Board of Sunmoon University (IRB No. SM-202206-030-2) and was prospectively registered in the Clinical Research Infor-mation Service (CRIS; KCT0008368).

Informed Consent Statement: Informed consent was obtained from all individual participants included in the study.

Data Availability Statement: The datasets generated and analyzed during the current study are not publicly available due to institutional policy but are available from the corresponding authors on reasonable request.

Conflicts of Interest: The authors declare no conflicts of interest.

Abbreviations

AR	Augmented reality
ARG	Augmented reality exercise group
PT	Physical therapist
PTG	Physical therapist exercise group
30CST	30-second chair stand test
TUG	Timed Up and Go
2MST	2-Minute Step Test
SRT	Sit-and-Reach Test
F8W	Figure 8 walking test
HGRt	Hand grip right side
HGLt	Hand grip left side

References

1. Ng, Y.L.; Ma, F.; Ho, F.K.; Ip, P.; Fu, K.W. Effectiveness of virtual and augmented reality-enhanced exercise on physical activity, psychological outcomes, and physical performance: A systematic review and meta-analysis of randomized controlled trials. *Comput. Hum. Behav.* **2019**, *99*, 278–291. [CrossRef]
2. Postolache, O.; Monge, J.; Alexandre, R.; Geman, O.; Jin, Y.; Postolache, G. Virtual reality and augmented reality technologies for smart physical rehabilitation. In *Advanced Systems for Biomedical Applications*; Springer: Cham, Switzerland, 2021; pp. 155–180.
3. Cary, F.; Postolache, O.; Girao, P.S. Kinect based system and serious game motivating approach for physiotherapy assessment and remote session monitoring. *Int. J. Smart Sens. Intell. Syst.* **2014**, *7*, 1. [CrossRef]
4. Daponte, P.; De Vito, L.; Riccio, M.; Sementa, C. Design and validation of a motion-tracking system for ROM measurements in home rehabilitation. *Measurement* **2014**, *55*, 82–96. [CrossRef]
5. Cavalcanti, V.C.; de Santana Ferreira, M.I.; Teichrieb, V.; Barioni, R.R.; Correia, W.F.M.; Da Gama, A.E.F. Usability and effects of text, image and audio feedback on exercise correction during augmented reality based motor rehabilitation. *Comput. Graph.* **2019**, *85*, 100–110. [CrossRef]
6. Viglialoro, R.M.; Condino, S.; Turini, G.; Carbone, M.; Ferrari, V.; Gesi, M. Review of the augmented reality systems for shoulder rehabilitation. *Information* **2019**, *10*, 154. [CrossRef]
7. Monteiro, A.M.; Rodrigues, S.; Matos, S.; Teixeira, J.E.; Barbosa, T.M.; Forte, P. The Effects of 32 Weeks of Multicomponent Training with Different Exercises Order in Elderly Women's Functional Fitness and Body Composition. *Medicina* **2022**, *58*, 628. [CrossRef]
8. Medrano-Ureña, M.D.R.; Ortega-Ruiz, R.; Benítez-Sillero, J.D.D. Physical fitness, exercise self-efficacy, and quality of life in adulthood: A systematic review. *Int. J. Environ. Res. Public Health* **2020**, *17*, 6343. [CrossRef]
9. Vanhees, L.; Lefevre, J.; Philippaerts, R.; Martens, M.; Huygens, W.; Troosters, T.; Beunen, G. How to assess physical activity? How to assess physical fitness? *Eur. J. Prev. Cardiol.* **2005**, *12*, 102–114. [CrossRef]
10. Gudlaugsson, J.; Gudnason, V.; Aspelund, T.; Siggeirsdottir, K.; Olafsdottir, A.S.; Jonsson, P.V.; Arngrimsson, S.A.; Harris, T.B.; Johannsson, E. Effects of a 6-month multimodal training intervention on retention of functional fitness in older adults: A randomized-controlled cross-over design. *Int. J. Behav. Nutr. Phys. Act.* **2012**, *9*, 107. [CrossRef]

11. Arrieta, H.; Rezola-Pardo, C.; Gil, J.; Kortajarena, M.; Zarrazquin, I.; Echeverria, I.; Mugica, I.; Limousin, M.; Rodriguez-Larrad, A.; Irazusta, J. Effects of an individualized and progressive multicomponent exercise program on blood pressure, cardiorespiratory fitness, and body composition in long-term care residents: Randomized controlled trial. *Geriatr. Nurs.* **2022**, *45*, 77–84. [CrossRef]
12. Comeras-Chueca, C.; Villalba-Heredia, L.; Perez-Lasierra, J.L.; Lozano-Berges, G.; Matute-Llorente, A.; Vicente-Rodriguez, G.; Casajus, J.A.; Gonzalez-Aguero, A. Effect of an Active Video Game Intervention Combined With Multicomponent Exercise for Cardiorespiratory Fitness in Children With Overweight and Obesity: Randomized Controlled Trial. *JMIR Serious Games* **2022**, *10*, e33782. [CrossRef] [PubMed]
13. Faul, F.; Erdfelder, E.; Buchner, A.; Lang, A.G. Statistical power analyses using G* Power 3.1: Tests for correlation and regression analyses. *Behav. Res. Methods* **2009**, *41*, 1149–1160. [CrossRef] [PubMed]
14. Kim, Y.M.; Kim, S.; Bae, J.; Kim, S.H.; Won, Y.J. Association between relative hand-grip strength and chronic cardiometabolic and musculoskeletal diseases in Koreans: A cross-sectional study. *Arch. Gerontol. Geriatr.* **2021**, *92*, 104181. [CrossRef] [PubMed]
15. Mehmet, H.; Yang, A.W.; Robinson, S.R. What is the optimal chair stand test protocol for older adults? A systematic review. *Disabil. Rehabil.* **2020**, *42*, 2828–2835. [CrossRef] [PubMed]
16. Podsiadlo, D.; Richardson, S. The timed "Up & Go": A test of basic functional mobility for frail elderly persons. *J. Am. Geriatr. Soc.* **1991**, *39*, 142–148. [CrossRef]
17. Bohannon, R.W.; Bubela, D.J.; Wang, Y.C.; Magasi, S.S.; Gershon, R.C. Six-minute walk test versus three-minute step test for measuring functional endurance (alternative measures of functional endurance). *J. Strength Cond. Res.* **2015**, *29*, 3240. [CrossRef]
18. Mayorga-Vega, D.; Merino-Marban, R.; Viciana, J. Criterion-related validity of sit-and-reach tests for estimating hamstring and lumbar extensibility: A meta-analysis. *J. Sports Sci. Med.* **2014**, *13*, 1.
19. Brach, J.S.; Perera, S.; Shuman, V.; Gil, A.B.; Kriska, A.; Nadkarni, N.K.; Rockette-Wagner, B.; Cham, R.; VanSwearingen, J.M. Effect of Timing and Coordination Training on Mobility and Physical Activity Among Community-Dwelling Older Adults: A Randomized Clinical Trial. *JAMA Netw. Open* **2022**, *5*, e2212921. [CrossRef]
20. Park, S.; Lee, S.; Lee, Y.; Lee, M. Normative Reference Values of Physical Fitness Levels in Koreans: Results from the National Fitness Award Project (2017–2019). *Exerc. Sci.* **2022**, *31*, 511–526. [CrossRef]
21. Celis-Morales, C.A.; Welsh, P.; Lyall, D.M.; Steell, L.; Petermann, F.; Anderson, J.; Iliodromiti, S.; Sillars, A.; Graham, N.; Mackay, D.F.; et al. Associations of grip strength with cardiovascular, respiratory, and cancer outcomes and all cause mortality: Prospective cohort study of half a million UK Biobank participants. *BMJ* **2018**, *361*, k1651. [CrossRef]
22. Pettersson-Pablo, P.; Nilsson, T.K.; Hurtig-Wennlöf, A. Handgrip strength reference intervals in Swedish, young, healthy adults—The LBA study. *Nutrition* **2023**, *105*, 111867. [CrossRef]
23. Petermann-Rocha, F.; Parra-Soto, S.; Cid, V.; Venegas, P.; Huidobro, A.; Ferreccio, C.; Celis-Morales, C. The association between walking pace and grip strength and all-cause mortality: A prospective analysis from the MAUCO cohort. *Maturitas* **2023**, *168*, 37–43. [CrossRef] [PubMed]
24. Ho, F.K.; Celis-Morales, C.A.; Petermann-Rocha, F.; Sillars, A.; Welsh, P.; Welsh, C.; Anderson, J.; Lyall, D.M.; Mackay, D.F.; Sattar, N.; et al. The association of grip strength with health outcomes does not differ if grip strength is used in absolute or relative terms: A prospective cohort study. *Age Ageing* **2019**, *48*, 684–691. [CrossRef]
25. Labott, B.K.; Bucht, H.; Morat, M.; Morat, T.; Donath, L. Effects of exercise training on handgrip strength in older adults: A meta-analytical review. *Gerontology* **2019**, *65*, 686–698. [CrossRef]
26. Kümmel, J.; Kramer, A.; Giboin, L.S.; Gruber, M. Specificity of balance training in healthy individuals: A systematic review and meta-analysis. *Sports Med.* **2016**, *46*, 1261–1271. [CrossRef]
27. Yang, C.M.; Hsieh, J.S.C.; Chen, Y.C.; Yang, S.Y.; Lin, H.C.K. Effects of Kinect exergames on balance training among community older adults: A randomized controlled trial. *Medicine* **2020**, *99*, e21228. [CrossRef]
28. Jovanov, E.; Wright, S.; Ganegoda, H. Development of an Automated 30 Second Chair Stand Test Using Smartwatch Application. In Proceedings of the 2019 41st Annual International Conference of the IEEE Engineering in Medicine and Biology Society (EMBC), Berlin, Germany, 23–27 July 2019; pp. 2474–2477.
29. Roongbenjawan, N.; Siriphorn, A. Accuracy of modified 30-s chair-stand test for predicting falls in older adults. *Ann. Phys. Rehabil. Med.* **2020**, *63*, 309–315. [CrossRef] [PubMed]
30. Jahanpeyma, P.; Kayhan Koçak, F.Ö.; Yıldırım, Y.; Şahin, S.; Şenuzun Aykar, F. Effects of the Otago exercise program on falls, balance, and physical performance in older nursing home residents with high fall risk: A randomized controlled trial. *Eur. Geriatr. Med.* **2021**, *12*, 107–115. [CrossRef]
31. Cho, K.H.; Bok, S.K.; Kim, Y.J.; Hwang, S.L. Effect of lower limb strength on falls and balance of the elderly. *Ann. Rehabil. Med.* **2012**, *36*, 386–393. [CrossRef] [PubMed]
32. Lee, Y.C.; Chang, S.F.; Kao, C.Y.; Tsai, H.C. Muscle strength, physical fitness, balance, and walking ability at risk of fall for prefrail older people. *BioMed Res. Int.* **2022**, *2022*, 4581126. [CrossRef]
33. Gil, M.J.V.; Gonzalez-Medina, G.; Lucena-Anton, D.; Perez-Cabezas, V.; Ruiz-Molinero, M.D.C.; Martín-Valero, R. Augmented reality in physical therapy: Systematic review and meta-analysis. *JMIR Serious Games* **2021**, *9*, e30985. [CrossRef]

34. Lee, J.; Yu, J.; Hong, J.; Lee, D.; Kim, J.; Kim, S. The Effect of Augmented Reality-Based Proprioceptive Training Program on Balance, Positioning Sensation and Flexibility in Healthy Young Adults: A Randomized Controlled Trial. *Healthcare* **2022**, *10*, 1202. [CrossRef]
35. Ku, J.; Kim, Y.J.; Cho, S.; Lim, T.; Lee, H.S.; Kang, Y.J. Three-dimensional augmented reality system for balance and mobility rehabilitation in the elderly: A randomized controlled trial. *Cyberpsychol. Behav. Soc. Netw.* **2019**, *22*, 132–141. [CrossRef]
36. Jaiswal, P.R.; Qureshi, I.; Phansopkar, P.A. Effectiveness of Mulligan's Two-Leg Rotation Versus Muscle Energy Technique in Subjects With Hamstring Tightness. *Cureus* **2022**, *14*, e28890. [CrossRef]
37. Kawano, M.M.; Ambar, G.; Oliveira, B.I.; Boer, M.C.; Cardoso, A.P.; Cardoso, J.R. Influence of the gastrocnemius muscle on the sit-and-reach test assessed by angular kinematic analysis. *Braz. J. Phys. Ther.* **2010**, *14*, 10–15. [CrossRef] [PubMed]
38. Kompal, R.; Ul-Islam, S.; Bhatti, U.; Tahir, K.; Noor, R.; Kashif, M. Effects of static stretch versus hold relax in improving flexibility of tight hamstrings. *Rehman J. Health Sci.* **2022**, *4*, 36–40.
39. Genç, H.; Poyraz İşleyen, T.; Tahmaz, T.; Aydoğan, Y.; Atılgan, E. Effectiveness of telerehabilitation on hamstring flexibility in healthy adults. *Phys. Educ. Theory Methodol.* **2022**, *22*, 229–236. [CrossRef]
40. Hong, J.; Kim, J.; Kim, S.W.; Kong, H.J. Effects of home-based tele-exercise on sarcopenia among community-dwelling elderly adults: Body composition and functional fitness. *Exp. Gerontol.* **2017**, *87*, 33–39. [CrossRef]
41. Haas, F.; Sweeney, G.; Pierre, A.; Plusch, T.; Whiteson, J. Validation of a 2 minute step test for assessing functional improvement. *Open J. Ther. Rehabil.* **2017**, *5*, 71. [CrossRef]
42. Srithawong, A.; Poncumhak, P.; Manoy, P.; Kumfu, S.; Promsrisuk, T.; Prasertsri, P.; Boonla, O. The optimal cutoff score of the 2-min step test and its association with physical fitness in type 2 diabetes mellitus. *J. Exerc. Rehabil.* **2022**, *18*, 214–221. [CrossRef]
43. Węgrzynowska-Teodorczyk, K.; Mozdzanowska, D.; Josiak, K.; Siennicka, A.; Nowakowska, K.; Banasiak, W.; Jankowska, E.A.; Ponikowski, P.; Woźniewski, M. Could the two-minute step test be an alternative to the six-minute walk test for patients with systolic heart failure? *Eur. J. Prev. Cardiol.* **2016**, *23*, 1307–1313. [CrossRef]
44. Pedrosa, R.; Holanda, G. Correlation between the walk, 2-minute step and TUG tests among hypertensive older women. *Braz. J. Phys. Ther.* **2009**, *13*, 252–256. [CrossRef]
45. Jeon, S.; Kim, J. Effects of augmented-reality-based exercise on muscle parameters, physical performance, and exercise self-efficacy for older adults. *Int. J. Environ. Res. Public Health* **2020**, *17*, 3260. [CrossRef] [PubMed]
46. Liguori, G.; American College of Sports Medicine. *ACSM's Guidelines for Exercise Testing and Prescription*; Lippincott Williams Wilkins: Philadelphia, PA, USA, 2020.
47. Nualyong, T.; Siriphorn, A. Accuracy of the figure of 8 walk test with and without dual-task to predict falls in older adults. *J. Bodyw. Mov. Ther.* **2022**, *30*, 69–75. [CrossRef] [PubMed]
48. Toledo-Peral, C.L.; Vega-Martínez, G.; Mercado-Gutiérrez, J.A.; Rodríguez-Reyes, G.; Vera-Hernández, A.; Leija-Salas, L.; Gutiérrez-Martínez, J. Virtual/Augmented Reality for Rehabilitation Applications Using Electromyography as Control/Biofeedback: Systematic Literature Review. *Electronics* **2022**, *11*, 2271. [CrossRef]
49. Osumi, M.; Inomata, K.; Inoue, Y.; Otake, Y.; Morioka, S.; Sumitani, M. Characteristics of phantom limb pain alleviated with virtual reality rehabilitation. *Pain Med.* **2019**, *20*, 1038–1046. [CrossRef]
50. Ambron, E.; Buxbaum, L.J.; Miller, A.; Stoll, H.; Kuchenbecker, K.J.; Coslett, H.B. Virtual reality treatment displaying the missing leg improves phantom limb pain: A small clinical trial. *Neurorehabilit. Neural Repair* **2021**, *35*, 1100–1111. [CrossRef]
51. Bandura, A. Social cognitive theory of mass communication. In *Media Effects*; Routledge: Abingdon, UK, 2009; pp. 110–140.
52. Kim, H.; Lee, H.J.; Cho, H.; Kim, E.; Hwang, J. Replacing self-efficacy in physical activity: Unconscious intervention of the AR game, Pokémon GO. *Sustainability* **2018**, *10*, 1971. [CrossRef]

Disclaimer/Publisher's Note: The statements, opinions and data contained in all publications are solely those of the individual author(s) and contributor(s) and not of MDPI and/or the editor(s). MDPI and/or the editor(s) disclaim responsibility for any injury to people or property resulting from any ideas, methods, instructions or products referred to in the content.

bioengineering

MDPI

Review

Muscle Fatigue in Dynamic Movement: Limitations and Challenges, Experimental Design, and New Research Horizons

Natalia Daniel [1], Jerzy Małachowski [2,*], Kamil Sybilski [2] and Michalina Błażkiewicz [3]

1 Institute of Rocket Technology and Mechatronics, Faculty of Mechatronics, Armament and Aviation, Military University of Technology, 00-908 Warsaw, Poland; natalia.daniel@wat.edu.pl
2 Institute of Mechanics & Computational Engineering, Faculty of Mechanical Engineering, Military University of Technology, 00-908 Warsaw, Poland; kamil.sybilski@wat.edu.pl
3 Faculty of Rehabilitation, Józef Piłsudski University of Physical Education in Warsaw, 00-968 Warsaw, Poland; michalina.blazkiewicz@awf.edu.pl
* Correspondence: jerzy.malachowski@wat.edu.pl

Abstract

Research on muscle fatigue during dynamic movement using surface electromyography (sEMG) constitutes a significant challenge within biomechanics. Despite a degree of standardization, measurements and their resultant findings continue to attract considerable debate, attributable to factors such as skin impedance, perspiration, and electrode displacement, as well as subjective fatigue perception. Further questions remain regarding signal normalization and the selection of appropriate analytical methodologies. Recent years have witnessed notable progress in dynamic fatigue research, highlighting the limitations of classical metrics (e.g., EMG Median Frequency) and introducing time–frequency methods, such as the wavelet transform (WT), which are better equipped to handle signal non-stationarity. Interest has also expanded to include non-linear metrics (e.g., entropy) and the analysis of multiple signals (EMG, accelerometers, fNIRS, EEG). The inherent complexity of conducting studies under conditions that approximate real-world sporting disciplines requires the consideration of the influence of various confounding factors. The judicious selection of relevant physical activities and the rigorous validation of the measurement apparatus are paramount for the accurate execution of the calculations. Current research is substantially predicated on artificial intelligence (AI) algorithms. The synergistic application of AI with wavelet transform, particularly in the decomposition and extraction of EMG signals, demonstrates efficacy in fatigue detection. Nevertheless, the full realization of these potential mandates requires further investigation into system generalization, the integration of data from multiple sensors, and the standardization of protocols, coupled with the establishment of publicly accessible datasets. This article delineates selected guidelines and challenges pertinent to the planning and execution of research on muscle fatigue in dynamic movement, focusing on activity selection, equipment validation, EMG signal analysis, and AI utilization.

Keywords: EMG; DWT; AI; dynamic movement

1. Introduction

Investigating muscle fatigue during dynamic movement remains a formidable challenge in biomechanics. Although surface electromyography (sEMG) has been used for decades to quantify muscle activity and fatigue [1,2], and despite efforts to standardize these procedures [3–5], the optimal methodological approach and interpretation of results

https://doi.org/10.3390/bioengineering13020248

remain subjects of significant debate [6]. Recently, numerous guidelines have emerged that aim to mitigate the limitations frequently encountered by researchers [7–9]. Critical confounding factors include the impedance of the electrode-skin, which compromises the signal fidelity [10], perspiration, motion artifacts (e.g., electrode displacement or detachment), and the anatomical variability of electrode placement sites [11]. These technical challenges are further compounded by physiological variables, such as subjective fatigue perception [12]. From a signal processing perspective, researchers face persistent questions regarding signal normalization strategies [13], the selection of appropriate computational methods, and the standardization of protocols across different sports activities [14].

Significant progress has recently been made in characterizing muscle fatigue during dynamic tasks. It is currently well established that classical spectral indices, such as Median Frequency (MDF) or Mean Frequency (MNF), provide useful but ultimately insufficient information for dynamic conditions. To address the inherent non-stationarity of these signals, time–frequency methods—including the Wavelet Transform (WT)—and adaptive techniques such as Empirical Mode Decomposition (EMD) and the Hilbert–Huang Transform (HHT) have been implemented to facilitate continuous fatigue monitoring [15,16]. Wavelet-based analyses have demonstrated superior stability compared to the Fast Fourier Transform (FFT) in long-duration dynamic experiments [15]. Despite these advantages, alternative methods continue to be explored. Interest has surged in non-linear metrics, such as entropy [17], and parameter assessment based on multiple signals, the combination of EMG measurements with accelerometers [18], Functional Near-Infrared Spectroscopy (fNIRS) or Electroencephalography (EEG) [19,20]. This methodological complexity offers opportunities to evaluate fatigue under conditions that approximate real-world performance, including the emerging study of fatigue within simulated Virtual Reality (VR) environments [21]. However, this field currently lacks a consistent framework, particularly regarding the quantitative application of Discrete Wavelet Transform (DWT) and the establishment of testing protocols. Even when computational methods are theoretically implemented, the artifacts and measurement errors inherent to various experimental setups frequently compromise the reliability of the data.

Given the increasing volume and complexity of data derived from these multi-modal setups, contemporary research increasingly relies on artificial intelligence (AI) to automate fatigue detection [22]. AI algorithms offer a robust mechanism to combine various signal inputs and identify subtle patterns of exhaustion that traditional statistical methods may overlook [23], and the successful deployment of AI in this context requires a fundamental understanding of the biophysical origin and the artifacts involved [24]. Consequently, substantial potential for advancement lies in the development of hybrid frameworks that integrate advanced signal decomposition (specifically time–frequency analysis) with AI classifiers. Such an approach is paramount for mitigating signal non-stationarity and facilitating the transition from post hoc analysis to continuous, real-time monitoring in dynamic environments.

Based on the authors' previous work addressing these methodological hurdles [25–27], this article presents selected guidance, formulated based on current guidelines and the own experiences gained from research on muscle fatigue in dynamic movement [25–27], this article presents selected guidelines, formulated based on current guidelines and the own experiences gained from research on muscle fatigue in dynamic movement. It focuses on the planning process for compiling research results in dynamic movement, addressing difficulties encountered at each stage. This article provides answers to common questions and offers approaches to solving specific problems. Individual issues are discussed based on research conducted on muscle fatigue in a homogeneous group of ergometer athletes. Ultimately, it demonstrates the potential to apply AI algorithms in conjunction with wavelet

https://doi.org/10.3390/bioengineering13020248

analysis, specifically highlighting research on fatigue in dynamic movement. Although the authors' previous work partially addressed the aforementioned issues, this article, informed by current guidelines, is considered to offer valuable insights for planning and executing experiments. The article is organized into sections, as illustrated in the schematic diagram in Figure 1.

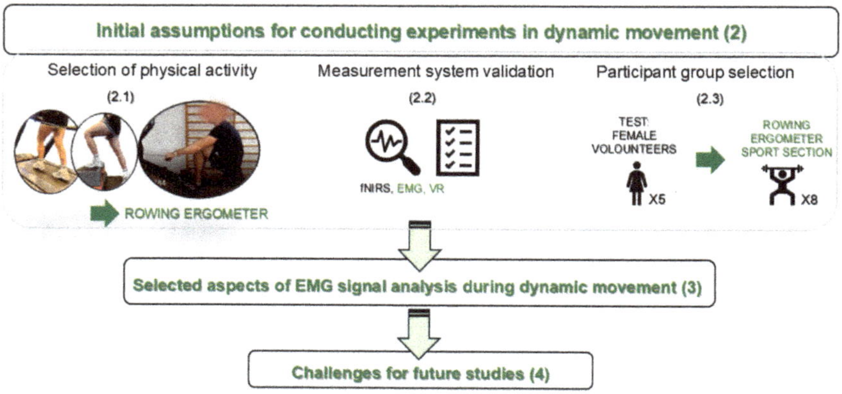

Figure 1. Schematic overview of the structure of the article and experimental workflow.

The remainder of this article follows the logical progression of the experimental workflow (Figure 1). Section 2 details the foundational study design, covering the rationale for selecting the rowing ergometer task (2.1), the validation of the multimodal measurement system (2.2), and the criteria for participant recruitment of participants (2.3). Subsequent sections address specific methodological strategies for EMG signal analysis during dynamic movement, concluding with a discussion of current limitations and challenges for future research.

Consequently, this manuscript is designed not only as a review of the existing literature but primarily as a methodological proposal and a practical guide. By integrating theoretical foundations with an evaluation of specific technical challenges (e.g., motion artifacts, sensor synchronization) identified during our pilot experiments, this paper aims to provide a robust framework for researchers. It addresses practical instructions for the evaluation of dynamic fatigue to facilitate the design of more reliable experimental setups in future studies.

2. Initial Assumptions for Conducting Experiments in Dynamic Movement

Investigating muscle fatigue during dynamic movement demands rigorous experimental planning. The key methodological prerequisites established prior to primary data acquisition include the selection of an appropriate physical activity, the definition of participant inclusion criteria, and the validation of measurement instrumentation. This section details these foundational steps and outlines the process that leads to the optimized research protocol.

2.1. Selection of Physical Activity

Selecting physical activity that simultaneously reflects dynamic conditions and allows for controlled repeatable measurement constitutes a key challenge in muscle fatigue research. The preliminary assumptions were based on the identification of the type of experiment and its intended duration. The objective was to select a movement modality that would involve specific muscle groups (in this case, the triceps suprae muscle was

https://doi.org/10.3390/bioengineering13020248

the target) and induce fatigue to a subjective maximum and enable fatigue induction to a subjective maximum. Although the current iteration of the protocol focuses on the lower extremities, the rowing ergometer inherently engages the upper body musculature. Consequently, extending the measurement setup to include upper extremity muscles constitutes a natural and planned expansion of this research framework to fully capture compensatory mechanisms during fatigue.

Based on the first work [25], two standardized tests were considered: the Astrand treadmill test, which involves running on a treadmill with gradually increasing incline, and the Astrand–Rhyming step test, which involves stepping onto a step at a set rhythm. One of the purposes of conducting these tests was to check how the measurement equipment would behave (EMG—Ultium EMG (Noraxon) 32 channels, sampling rate 4000 Hz—for recording muscle electrical signals; fNIRS—OctaMon M (Artinis)—measuring changes in muscle oxygenation, two receivers, and eight transmitters using near-infrared light) during long-term dynamic fatigue tests. These tests were chosen due to activation of the triceps suprae muscle and the possibility of assessing maximum fatigue using the Borg scale and variable load.

However, the pilot tests revealed significant methodological limitations. Despite strict adherence to the skin preparation guidelines (specifically, hair removal and thorough cleaning with an alcohol-ether mixture to minimize impedance), electrode displacement and detachment were frequent. Furthermore, the high-impact mechanics inherent in running and stepping introduced substantial motion artifacts into both the sEMG and fNIRS signal streams. These findings required a transition to a dynamic movement task that minimizes measurement disturbances while maintaining biomechanical familiarity for athletes.

Consequently, the use of a rowing ergometer was decided upon. This activity, due to its stationary nature and cyclical movements, facilitates more controlled monitoring of changes in muscle activity and mitigates variability arising from uncontrolled posture and technique, making fatigue measurement more repeatable and reliable. Furthermore, the selection of the ergometer was influenced by the intention to incorporate VR elements into the investigation [28]. Numerous dynamic sports had to be excluded due to the potential health risks to the participants during the measurement, which was conducted in a completely detached environment.

The duration of the experimental trials was not rigidly standardized; instead, the protocol required continuation until the participant reached a state of voluntary exhaustion. This approach was critical, as imposing a fixed time limit would have rendered the results unreliable by failing to capture the true threshold of neuromuscular fatigue. These tests were chosen due to the activation of the specific muscles and the possibility of assessing maximum fatigue using the Borg scale and variable load. Furthermore, in rowing ergometer trials, variable damper resistance and continuous Heart Rate (HR) monitoring were used to safely induce and verify this maximal physiological effort.

2.2. Measurement System Validation

During the selection of physical activity, a validation of the measurement instrumentation was carried out to assess its operation and dependability under dynamic conditions. The initial configuration consisted of the EMG Ultium EMG systems (Noraxon), referenced in the previous chapter (2.1), to record muscle electrical signals and the OctaMon M (Artinis) to measure changes in muscle oxygenation (fNIRS). Both were mounted on the triceps suprae muscle, while movement was analyzed concurrently using a camera. Throughout the dynamic tests (treadmill, step), several equipment-related challenges were noted:

https://doi.org/10.3390/bioengineering13020248

- Due to differences in running technique and significant changes associated with leg movement, electrodes are often moved and detached.
- Despite adhering to the literature guidelines, the problem remains unresolved.
- Fatigue tests, although defined by detailed experimental conditions, introduce many artifacts in the readings of both sensors.
- The color of the skin, its thickness and the individual anatomical differences cause disturbances in the fNIRS sensors.
- Synchronizing two independent sensors requires additional time, which affects the quality of research and the well-being of participants.
- The Borg scale provides a good starting point to determine the level of fatigue.

The choice of the rowing ergometer contributed to partial resolution of these issues. Data synchronization between EMG and fNIRS streams was performed manually during the post-processing phase. The application of integrated wearable sensors [29] may eliminate the challenges of manual integration, and this solution may be beneficial in future research to automate the data acquisition process. The more predictable and restricted range of movement on the ergometer reduced the problem of electrode displacement and mitigated the number of movement artifacts in the signals. Furthermore, the controlled indoor laboratory environment inherent to ergometer training minimizes ambient light interference (e.g., direct sunlight), effectively protecting the fNIRS signal from external optical noise. However, it should be noted that the planned expansion of research to include a greater number of sensors, such as an fNIRS unit positioned on the participant's head, may introduce conflicts with VR goggles, particularly in the context of excessive perspiration during exertion, which has been identified as a potential limitation.

2.3. Participant Group Selection

To increase the reliability of the results, particularly in the context of assessing fatigue during effort specific to a given sport discipline, the aim is to study homogeneous groups, such as athletes regularly training a specific discipline. To ensure the reliability of the data, specific measures were implemented to address physiological variability. Inter-personal variability was minimized by recruiting a homogeneous cohort of elite athletes. Furthermore, to mitigate intra-personal variability, participants adhered to a strict preparatory protocol (including dietary and sleep standardization). Detailed descriptions of these inclusion criteria and variability control procedures are provided in our previous experimental works [25–27]. One of the most important factors is that the group is selected in a way that ensures that it is representative. Statistical planning prior to the test is also crucial. However, achieving the assumed sample size is often challenging due to the highly demanding nature of research involving dynamic movement. Furthermore, while such homogeneity improves sport-specific analysis, it introduces a potential bias when developing universal AI frameworks, creating the risk of overfitting to the unique biometric patterns of elite athletes. Therefore, these aspects should be carefully considered during the experimental design phase to ensure an appropriate balance between population specificity and model generalizability.

The Borg scale was used to assess the subjective level of fatigue. It constitutes a valuable reference point, allowing for individualization of the test completion time or assessment of fatigue progression.

Psychological and behavioral factors are also significant. The appropriate preparation for the test, the proper hydration of the body, and the motivation of the participants can influence their performance and subjective sensations, which should be considered in the research protocol.

https://doi.org/10.3390/bioengineering13020248

Experimental design should be grounded in a priori statistical power and effect size analysis (e.g., using G*Power 3.1 software), which often indicates a requirement for large sample sizes (N > 50) to maximize statistical power. However, in dynamic research involving elite athletes, such numbers are often logistically unfeasible due to population scarcity. The methodological insights presented in this paper are derived from our previous experimental iterations that involved smaller, highly homogeneous cohorts (e.g., N = 5 to N = 8). Therefore, the focus of this work is to synthesize these experiences into a robust framework that assists researchers in navigating these constraints, prioritizing high data quality and group homogeneity when large-scale recruitment is not possible. From a machine learning perspective, the definition of an adequate sample size can shift from the number of participants to the total number of movement cycles. Through signal segmentation, where each specific motion constitutes a single training instance, even smaller cohorts yield thousands of data vectors. This volume is typically sufficient for feature-based classifiers, while scenarios involving data-demanding Deep Learning models may necessitate the use of Data Augmentation or Transfer Learning techniques. Consequently, the assessment of dataset sufficiency in such frameworks often relies on learning curve analysis rather than classical a priori power estimation.

2.4. Summary

The factors outlined above regarding the selection of physical activity, the validation of the measurement apparatus, and the recruitment of the participant group, informed by the findings of the preliminary studies, established the basis for the development of a novel and detailed measurement protocol. The intent behind these endeavors was to establish the most objective and reproducible conditions feasible for investigating the phenomenon of muscle fatigue in dynamic movement while simultaneously mitigating the impact of confounding factors. The workflow of the research is depicted schematically in Figure 2.

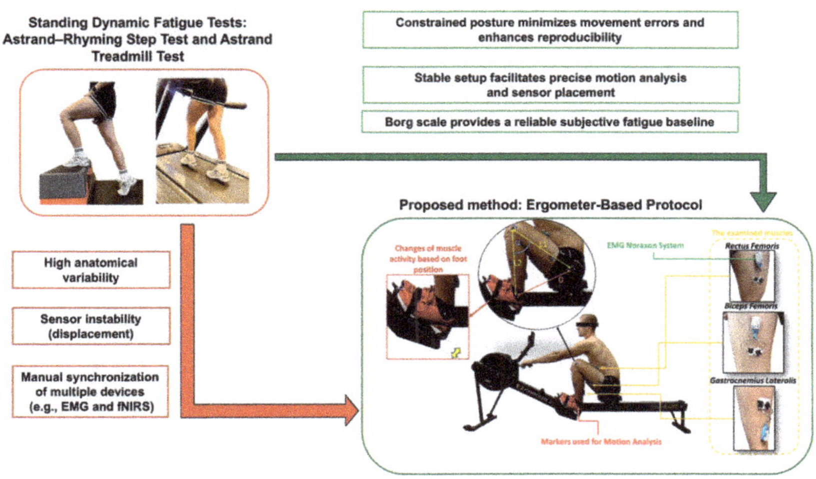

Figure 2. Dynamic fatigue experiments that activate the muscles of the lower extremities are derived from many standing exercises, and the proposed method is based on studies by the previous author [27]—research workflow.

EMG signal analysis is a commonly used technique to study muscle physiology and investigate muscle fatigue [7,30], which provides the ability to monitor changes associated with progressive exertion. The key parameters used in this approach are the MNF and MDF of the EMG signal power spectrum, which are recognized as valuable indicators

https://doi.org/10.3390/bioengineering13020248

of fatigue [31–33]. Numerous investigations have established that a critical indicator of developing fatigue, observable in spectral analysis, is the typical downward shift in the EMG signal power spectrum toward lower frequencies, which correlates with changes in the conduction velocity of action potentials in muscle fibers. This occurrence is characterized by a systematic decrease in both calculated MDF and MNF values as the duration of fatiguing contractions lengthens [34,35]. A negative linear regression slope coefficient for changes in MDF over time is considered an effective quantitative indicator that allows assessing the onset and progression of muscle fatigue [36–38].

The EMG signal is widely considered a non-stationary signal. This means that its general characteristics can change dynamically during the recording of muscle activity. This occurs due to complex physiological phenomena, such as muscle fatigue progression, changes in motor unit recruitment, or fluctuations in the level of generated force. This non-stationarity of the EMG signal is a crucial aspect that must be considered when selecting analysis methods to ensure the reliability and validity of the obtained results. Applying methods that assume stationarity can lead to erroneous interpretations.

3. Selected Aspects of EMG Signal Analysis During Dynamic Movement

3.1. Standard Spectral Analysis Under Static Conditions

The selection of an appropriate methodology for the analysis of EMG signals, a key factor in muscle fatigue investigations, is dictated by the characteristics of the movement under examination and the attributes of the signal itself. Considering that this analysis, which assesses parameters such as MDF and MNF, is concerned with the relative power distribution in the frequency domain rather than the signal's absolute amplitude, EMG normalization is not necessary for this kind of evaluation. As highlighted in a review [13], in frequency analysis, the amplitude of the EMG signal is not critical, and normalization of the EMG is not required. Popular analysis techniques include: FFT, Short-Time Fourier Transform (STFT), and WT [31,39,40]. In the case of static (isometric) movement, where the EMG signal can be treated as quasi-stationary in short time windows, it is often sufficient to apply classical methods of spectrum estimation, such as FFT [41], for the calculation and tracking of changes in the parameters described above, such as MDF/MNF. Referring to the previous chapter, the EMG signal exhibits a non-stationary character [42], which can limit the effectiveness of the classical Fourier transform [43], especially during longer analyses or when fatigue increases.

3.2. Time–Frequency Analysis for Dynamic Movements

Analysis of dynamic movement, which generates a highly non-stationary EMG signal resulting from complex changes in muscle activity and movement kinematics, necessitates the use of more advanced time–frequency analysis techniques capable of precisely tracking spectral changes over time. For this reason, STFT and WT are often considered to be more appropriate analysis methods under such conditions [44]. STFT analyzes the signal in sliding time windows. However, its use involves a fundamental compromise between temporal and frequency resolution, depending on the width of the chosen analysis window [45]. On the contrary, the wavelet transform is indicated as a method that offers potentially superior accuracy and adaptive time–frequency resolution (better time resolution for high frequencies and better frequency resolution for low frequencies), which is particularly advantageous when analyzing nonstationary signals and dynamic contractions [41,45,46]. Within this family of transforms, the DWT is often preferred over the continuous wavelet transform (CWT) due to its reduced redundancy of information and computational efficiency [43,47]. DWT implementation typically employs filter banks to decompose the signal

https://doi.org/10.3390/bioengineering13020248

into approximation and detail coefficients on different frequency scales [47,48], making it an effective tool for EMG analysis in the context of fatigue assessment.

Before proceeding to the subsequent stages of research development and employing additional measurements or computational methods, it is beneficial to familiarize oneself with the differences in calculations between static and dynamic movement. Furthermore, if the research is planned to be extended to include other devices (for example, fNIRS, EEG, or the implementation of VR), it is also necessary to become familiar with the methods of signal filtration, artifact removal, segmentation, and classification.

4. Challenges for Future Studies on Fatigue During Dynamic Movement

4.1. Integrated Research Framework and Preprocessing

The aspects related to the planning of research on fatigue in dynamic movement, as well as the acquisition and processing of the EMG signal, have been discussed above. These are the main aspects that should be analyzed before beginning research using artificial intelligence tools. Figure 3 presents a schematic diagram of the research procedure for muscle fatigue in dynamic movement.

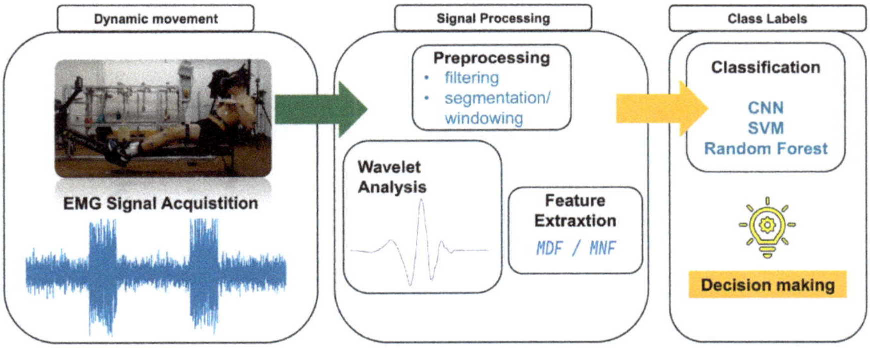

Figure 3. Schematic diagram of the research procedure for fatigue in dynamic movement.

Within the preliminary processing stage (preprocessing), raw EMG signals are subjected to filtration to eliminate noise and interference outside the physiological frequency band. For this purpose, a Butterworth filter is often employed, configured as a band-pass filter, for instance, with cutoff frequencies of 20 Hz and 500 Hz, which permits the removal of both low-frequency and high-frequency interference. Following the initial preparation of the signal, the primary analysis phase focuses on time–frequency methods, with particular emphasis on WT, often in its discrete form. The efficacy of WT/DWT in analyzing non-stationary EMG signals, particularly those recorded during dynamic movements, justifies its selection. DWT-based analysis involves decomposing the signal into distinct frequency levels using a chosen mother wavelet, such as Daubechies4 (db4), which yields approximation and detail coefficients. Subsequently, the frequency parameters, such as MDF, are calculated from these coefficients or the reconstructed signal, serving as a quantitative measure of changes in muscle fatigue during exercise. Ultimately, the features thus extracted, which condense the information within the EMG signal, are fed into AI algorithms, including both conventional machine learning (ML) methods and deep learning (DL) models. These algorithms undertake tasks such as pattern recognition, classification (e.g., fatigue status), or prediction, leveraging the potential synergy between advanced feature extraction techniques (like WT) and the interpretative capabilities of AI.

AI, particularly ML and DL, enables automatic analysis of complex signals, identification of patterns indicating fatigue that are difficult to recognize, and the construction of

https://doi.org/10.3390/bioengineering13020248

predictive models [49]. Various AI/ML/DL algorithms are used in this area; AI models allow the learning of complex patterns from features extracted from EMG data, providing an objective data-driven evaluation of fatigue [7].

Given the scientific need for a quantitative assessment of fatigue in dynamic movement, it was decided to further explore the directions of research on fatigue using AI techniques with wavelet analysis. This focuses on the role of wavelet analysis in feature extraction with the application of AI models (e.g., SVM, CNN, Random Forest) [50,51]. This is intended to define learning and validation strategies, comparing different methodologies and their limitations.

4.2. Wavelet-Based Feature Extraction Strategies

For each DWT level, the signal is decomposed into approximation coefficients (cA), which reflect the low-frequency constituents whose amplitude is associated with fatigue development, and detail coefficients (cD), representing the high-frequency constituents. The approximation and detail coefficients themselves, obtained at various levels of the DWT, can serve as feature vectors [52]. These vectors capture the energy distribution of the signal on various time–frequency scales. PCA can be applied to these coefficient vectors for dimensionality reduction [53].

The use of wavelet coefficients primarily serves to segment the EMG signal into more meaningful localized time–frequency components before the application of feature calculations (statistical, spectral, etc.) or their use as input data for machine learning models [52]. The primary advantage lies in the decomposition, rather than the direct use of raw coefficient values in all cases. Many studies extract statistical characteristics from the coefficients [54] or calculate the MDF based on these coefficients [27], rather than using the coefficients themselves as the final characteristic vector. Even when the coefficients are used more directly [53], they represent the energy distribution of the signal in specific bands. This pattern suggests that the primary role of WT here is effective signal decomposition and localization, enabling more significant feature extraction in appropriate sub-bands associated with fatigue.

4.3. Optimization of the WT-AI Synergy

The synergy between WT and AI in EMG signal analysis is particularly effective when WT is applied not only for general feature extraction but also as a tool for preliminary signal processing. This encompasses both effective noise removal, which is crucial given the susceptibility of EMG to interference, and the potential enhancement or isolation of specific muscle fatigue biomarkers [52]. For example, advanced wavelet-based denoising techniques, such as improved thresholding methods [49], can help preserve the spectral changes associated with fatigue. Similarly, the decomposition of the signal into precisely selected levels or the application of specific decomposition methods, such as ICEEMDAN [23], allows the isolation of frequencies significant to the fatigue process. Studies achieving the highest classification effectiveness often rely on specific optimizations of the wavelet stage, such as selecting decomposition levels or wavelet coefficients associated with fatigue [52]. This indicates that success in this field is largely due to the skillful use of WT for signal preparation, filtering out interference and highlighting fatigue indicators, which in turn can enable machine learning models to perform more effective and accurate classification. Consequently, numerous studies have demonstrated the effectiveness of an approach in which WT is employed in the preliminary processing stage (denoising) and/or for feature extraction, followed by the AI algorithm performing the classification of the fatigue state.

References can be found in the literature on the utilization of wavelet coefficients or signal energy in specific wavelet bands as input features for classifiers such as SVM or

https://doi.org/10.3390/bioengineering13020248

ANN [51]. These results indicate that the optimal AI-WT framework appears to be task-dependent. Different dynamic movements engage different patterns of muscle activation, velocities, and types of contraction. This suggests that tailoring the extraction of WT features and the selection of AI models to the specific biomechanics and anticipated EMG characteristics of a given dynamic task is crucial in muscle fatigue research.

4.4. Current State-of-the-Art: A Comparative Review

The ability of wavelet transform to precisely analyze the time–frequency characteristics of non-stationary EMG signals, in combination with the capabilities of AI/ML/DNN in the field of advanced pattern recognition and automatic feature learning, has led to significant progress in this area. A limited number of studies address the topic of fatigue itself, as evident from the above description. However, considering the topic more broadly, not limiting oneself to the dynamic movement aspect alone, the synergistic combination of AI, ML, and deep neural networks (DNN) with WT constitutes a powerful tool in the advanced analysis and processing of EMG signals.

To present the area of scientific research concerning the application of AI/ML/DNN in conjunction with wavelet transform for the analysis and processing of EMG signals, studies presenting selected research methodologies are compiled in Table 1.

Table 1. Comparative studies on the use of AI methods with wavelet transform.

Source	Technique	AI Methods	Application	Results
[55]	DWT	Intelligent dynamic bit allocation scheme implemented using a Kohonen layer (neural network)	Data compression, noise reduction	It has been demonstrated that the compression performance of the EMG signal is superior compared to standard wavelet algorithms, minimizing distortions at a given compression ratio.
[56]	WT	ANN—MLP and GRNN (input comprises the coefficients of the auto-regressive signal model after WT)	Muscle type classification, feature extraction	The high effectiveness in classifying muscle types was confirmed, and it was demonstrated that preliminary processing of EMG signals using the wavelet transform significantly improves the results.
[57]	DWT	Random Forest, KNN, Decision Tree	Hand movement classification, feature extraction	The effectiveness of different feature extractors and classifiers was compared, identifying combinations that produced good results; DWT features showed competitive effectiveness.
[58]	WPT (wavelet packet transform)	BPNN SVM GA-SVM	Muscle fatigue, muscle activity classification	Identification of muscle fatigue using the GA-SVM classifier, which was more accurate than other approaches.
[59]	DWT	Feed Forward Back Propagation Neural Network (FFBPNN) (ANN)	Hand movement classification, feature extraction	High accuracy was achieved in the classification of hand movements using DWT features and a neural network, indicating the effectiveness of the selected decomposition level.

https://doi.org/10.3390/bioengineering13020248

Table 1. *Cont.*

Source	Technique	AI Methods	Application	Results
[60]	CWT	Deep neural networks (ConvNets)	Gesture recognition, Feature extraction (automatic)	Transfer learning was shown to systematically and significantly improve the performance of deep neural networks in EMG gesture recognition, particularly for CWT models.
[61]	DWT	Adaptive Neuro-Fuzzy Inference System (ANFIS).	EMG signal recognition (for prosthetics)	High accuracy in the recognition of EMG signals was achieved, indicating the potential to combine DWT and ANFIS as a control signal for prosthetics.
[62]	DWT	KNN	Recognition of hand movements.	A practical and computationally lightweight multilevel feature extraction method (TP-DWT) was proposed for sEMG signals, which allowed for achieving high accuracy to be achieved in hand movement recognition.
[63]	CWT (cumulative scalograms)	XMANet (Cross-layer Mutual Attention Learning Network) with different CNNs	Gesture recognition, advanced feature extraction	A novel network architecture (XMANet) with attention mechanisms was proposed, which consistently improves the performance of gesture recognition based on CWT scalograms.

This mini-review highlights the main areas of research. They focus primarily on data compression and noise reduction in EMG signals, as well as muscle type classification, hand movement classification, gesture recognition, and EMG signal recognition for prosthetic control. As can be observed, the classification of hand movements and gesture recognition constitutes a significant area of these studies. Research concerning the detection of muscle fatigue and the classification of muscle activity represents a smaller proportion, indicating the need for further development in this thematic area.

5. Conclusions

Although the combination of wavelet analysis of EMG signals with machine learning methods represents a promising strategy for detecting muscle fatigue, fully exploiting its potential requires further research focused on several key challenges. A significant limitation of current systems is their limited generalizability beyond specific training conditions, which highlights the need to develop domain adaptation techniques and implement rigorous validation protocols that cover diverse populations and movement tasks. Furthermore, integrating EMG data with information from other physiological sensors, such as NIRS and EEG, within an approach based on multimodal fusion and machine learning, can significantly improve the complexity and reliability of fatigue assessment. Standardization efforts are equally fundamental, including the development of protocols for fatigue induction, statistical methods improvement, data acquisition, criteria for selecting mother wavelets, feature extraction methods and validation, as well as the creation of publicly available reference datasets. Such a harmonization will enable a reliable comparison of the results of different studies and will accelerate the translation of the developed technologies into practical applications. Addressing these challenges is crucial to further progress and the

https://doi.org/10.3390/bioengineering13020248

practical implementation of fatigue monitoring systems that use EMG, wavelet analysis, and machine learning. Ultimately, the refinement of this methodology paves the way for applications transcending the realm of competitive sports, finding critical utility in clinical rehabilitation and smart assistive technologies. Such a transition to real-time capability will empower future biofeedback systems to dynamically respond to muscle status, thereby ensuring safety and efficacy in both athletic training and therapeutic recovery.

Author Contributions: Conceptualization, methodology, investigation, formal analysis, data curation, visualization, writing—original draft—N.D.; methodology, project administration, resources, writing—review and editing—K.S. and J.M.; methodology, writing—review and editing—N.D., K.S., J.M. and M.B.; methodology, supervision, visualization, writing—review and editing—J.M. All authors have read and agreed to the published version of the manuscript.

Funding: The authors would like to express their sincere thanks to the Military University of Technology for its financial support (project no. UGB 103(1)/2026). This project is part of the university's internal grant system designed to support the statutory research activities of its employees.

Institutional Review Board Statement: This study received approval from the Ethics Committee of the Warsaw University of Life Sciences (number 19/2022).

Informed Consent Statement: Informed consent was obtained from all subjects involved in this study.

Data Availability Statement: The original contributions presented in this study are included in the article. Further inquiries can be directed to the corresponding author.

Acknowledgments: The authors would like to thank all the volunteers who agreed to participate in this study and the institution that provided all the resources required for this study.

Conflicts of Interest: The authors declare no conflicts of interest. The funders had no role in the design of the study; in the collection, analyses, or interpretation of data; in the writing of the manuscript; or in the decision to publish the results.

Abbreviations

The following abbreviations are used in this manuscript:

sEMG	Surface Electromyography
WT	Wavelet Transform
fNIRS	Functional Near-Infrared Spectroscopy
EEG	Electroencephalography
AI	Artificial Intelligence
EMG	Electromyography
MDF	Median Frequency
MNF	Mean Frequency
EMD	Empirical Mode Decomposition
HHT	Hilbert–Huang Transform
FFT	Fast Fourier Transform
VR	Virtual Reality
DWT	Discrete Wavelet Transform
CWT	Continuous Wavelet Transform
STFT	Short-Time Fourier Transform
db4	Daubechies 4
ML	Machine Learning
DL	Deep Learning
SVM	Support Vector Machine
CNN	Convolutional Neural Network
PCA	Principal Component Analysis
ANN	Artificial Neural Network

https://doi.org/10.3390/bioengineering13020248

DNN	Deep Neural Network
MLP	Multilayer Perceptron
GRNN	General Regression Neural Network
GA-SVM	Genetic Algorithm-based SVM
FFBPNN	Feed-Forward Back-Propagation Neural Network
ANFIS	Adaptive Neuro-Fuzzy Inference System
KNN	K-Nearest Neighbors
XMANet	Cross-layer Mutual Attention Learning Network

References

1. Viitasalo, J.H.T.; Komi, P.V. Signal Characteristics of EMG during Fatigue. *Eur. J. Appl. Physiol. Occup. Physiol.* **1977**, *37*, 111–121. [CrossRef]
2. Potvin, J.R.; Bent, L.R. A Validation of Techniques Using Surface EMG Signals from Dynamic Contractions to Quantify Muscle Fatigue during Repetitive Tasks. *J. Electromyogr. Kinesiol.* **1997**, *7*, 131–139. [CrossRef]
3. Ball, N.; Scurr, J. Electromyography Normalization Methods for High-Velocity Muscle Actions: Review and Recommendations. *J. Appl. Biomech.* **2013**, *29*, 600–608. [CrossRef]
4. Cifrek, M.; Medved, V.; Tonković, S.; Ostojić, S. Surface EMG Based Muscle Fatigue Evaluation in Biomechanics. *Clin. Biomech.* **2009**, *24*, 327–340. [CrossRef] [PubMed]
5. Tankisi, H.; Burke, D.; Cui, L.; de Carvalho, M.; Kuwabara, S.; Nandedkar, S.D.; Rutkove, S.; Stålberg, E.; van Putten, M.J.A.M.; Fuglsang-Frederiksen, A. Standards of Instrumentation of EMG. *Clin. Neurophysiol.* **2020**, *131*, 243–258. [CrossRef]
6. Turker, K.S. Electromyography: Some Methodological Problems and Issues. *Phys. Ther.* **1993**, *73*, 698–710. [CrossRef] [PubMed]
7. Sun, J.; Liu, G.; Sun, Y.; Lin, K.; Zhou, Z.; Cai, J. Application of Surface Electromyography in Exercise Fatigue: A Review. *Front. Syst. Neurosci.* **2022**, *16*, 893275. [CrossRef]
8. Chuang, T.D.; Acker, S.M. Comparing Functional Dynamic Normalization Methods to Maximal Voluntary Isometric Contractions for Lower Limb EMG from Walking, Cycling and Running. *J. Electromyogr. Kinesiol.* **2019**, *44*, 86–93. [CrossRef]
9. Ball, N.; Scurr, J. An Assessment of the Reliability and Standardisation of Tests Used to Elicit Reference Muscular Actions for Electromyographical Normalisation. *J. Electromyogr. Kinesiol.* **2010**, *20*, 81–88. [CrossRef]
10. Sae-lim, W.; Phukpattaranont, P.; Thongpull, K. Effect of Electrode Skin Impedance on Electromyography Signal Quality. In *Proceedings of the 2018 15th International Conference on Electrical Engineering/Electronics, Computer, Telecommunications and Information Technology (ECTI-CON)*; IEEE: Chiang Rai, Thailand, 2018; pp. 748–751. [CrossRef]
11. Hung, G.; Pallis, J. *Biomedical Engineering Principles in Sports*; Springer: Berlin/Heidelberg, Germany, 2004; Volume 1; ISBN 978-1-4613-4706-4.
12. Williams, N. The Borg Rating of Perceived Exertion (RPE) Scale. *Occup. Med.* **2017**, *67*, 404–405. [CrossRef]
13. Halaki, M.; Ginn, K. Normalization of EMG Signals: To Normalize or Not to Normalize and What to Normalize To? In *Computational Intelligence in Electromyography Analysis*; Naik, G.R., Ed.; IntechOpen: Rijeka, Croatia, 2012.
14. Gates, D.H.; Dingwell, J.B. The Effects of Muscle Fatigue and Movement Height on Movement Stability and Variability. *Exp. Brain Res.* **2011**, *209*, 525–536. [CrossRef]
15. Chowdhury, S.K.; Nimbarte, A.D. Comparison of Fourier and Wavelet Analysis for Fatigue Assessment during Repetitive Dynamic Exertion. *J. Electromyogr. Kinesiol.* **2015**, *25*, 205–213. [CrossRef]
16. Jero, S.E.; Ramakrishnan, S. Analysis of Muscle Fatigue Conditions in Surface EMG Signal with A Novel Hilbert Marginal Spectrum Entropy Method. In *Proceedings of the 2019 41st Annual International Conference of the IEEE Engineering in Medicine and Biology Society (EMBC)*; IEEE: Piscataway, NJ, USA, 2019; Volume 2019, pp. 2675–2678. [CrossRef]
17. Li, Z.; Zhang, B.; Wang, H.; Gouda, M.A. Surface Electromyography Monitoring of Muscle Changes in Male Basketball Players During Isotonic Training. *Sensors* **2025**, *25*, 1355. [CrossRef]
18. Fougner, A.; Scheme, E.; Chan, A.D.C.; Englehart, K.; Stavdahl, Ø. Resolving the Limb Position Effect in Myoelectric Pattern Recognition. *IEEE Trans. Neural Syst. Rehabil. Eng.* **2011**, *19*, 644–651. [CrossRef]
19. Shakya, S.; Ranjitkar, B. Forearm Bio-Medical Signal Processing. *Int. J. Eng. Technol.* **2024**, *2*, 49–59. [CrossRef]
20. Song, J.; Choi, Y.S.; Lee, S.; Park, D.; Park, J. Changes in Muscle Oxygenation and Activity during Cumulative Isometric Muscle Contraction: New Insight into Muscle Fatigue. *Front. Physiol.* **2025**, *16*, 1559893. [CrossRef]
21. Jayarathne, M.; Wickramanayake, D.; Afsheenjinan, A.; Ranaweera, R.; Weerasingha, V. EMG Based Biofeedback System Using a Virtual Reality Method. In *Proceedings of the 2015 IEEE 10th International Conference on Industrial and Information Systems (ICIIS)*; IEEE: Peradeniya, Sri Lanka, 2015; pp. 111–116. [CrossRef]
22. Subasi, A.; Kiymik, M.K. Muscle Fatigue Detection in EMG Using Time-Frequency Methods, ICA and Neural Networks. *J. Med. Syst.* **2010**, *34*, 777–785. [CrossRef] [PubMed]

https://doi.org/10.3390/bioengineering13020248

23. Sun, J.; Zhang, C.; Liu, G.; Cui, W.; Sun, Y.; Zhang, C. Detecting Muscle Fatigue during Lower Limb Isometric Contractions Tasks: A Machine Learning Approach. *Front. Physiol.* **2025**, *16*, 1547257. [CrossRef] [PubMed]
24. Reaz, M.B.I.; Hussain, M.S.; Mohd-Yasin, F. Techniques of EMG Signal Analysis: Detection, Processing, Classification and Applications. *Biol. Proced. Online* **2006**, *8*, 11–35. [CrossRef]
25. Daniel, N.; Sybilski, K.; Kaczmarek, W.; Siemiaszko, D.; Małachowski, J. Relationship between EMG and FNIRS during Dynamic Movements. *Sensors* **2023**, *23*, 5004. [CrossRef]
26. Daniel, N.; Małachowski, J. Wavelet Analysis of the EMG Signal to Assess Muscle Fatigue in the Lower Extremities during Symmetric Movement on a Rowing Ergometer. *Acta Bioeng. Biomech.* **2023**, *25*, 15–27. [CrossRef]
27. Daniel, N.; Małachowski, J.; Sybilski, K.; Siemiaszko, D. Quantitative Assessment of Muscle Fatigue during Rowing Ergometer Exercise Using Wavelet Analysis of Surface Electromyography (SEMG). *Front. Bioeng. Biotechnol.* **2024**, *12*, 1344239. [CrossRef]
28. Daniel, N.; Małachowski, J.; Sybilski, K.; Siemiaszko, D.; Klicki, K. Research on the Influence of Virtual Reality on Muscle Fatigue during Rowing Ergometer Exercise—Pilot Study. *PLoS ONE* **2026**, *21*, e0342166. [CrossRef] [PubMed]
29. Di Giminiani, R.; Cardinale, M.; Ferrari, M.; Quaresima, V. Validation of Fabric-Based Thigh-Wearable EMG Sensors and Oximetry for Monitoring Quadricep Activity during Strength and Endurance Exercises. *Sensors* **2020**, *20*, 4664. [CrossRef]
30. González-Izal, M.; Malanda, A.; Gorostiaga, E.; Izquierdo, M. Electromyographic Models to Assess Muscle Fatigue. *J. Electromyogr. Kinesiol.* **2012**, *22*, 501–512. [CrossRef] [PubMed]
31. Phinyomark, A.; Thongpanja, S.; Hu, H.; Phukpattaranont, P.; Limsakul, C. The Usefulness of Mean and Median Frequencies in Electromyography Analysis. In *Computational Intelligence in Electromyography Analysis—A Perspective on Current Applications and Future Challenges*; IntechOpen: London, UK, 2012. [CrossRef]
32. Kim, C.-B.; Park, C.-H.; Kim, C.-H.; Lee, H.-S.; Kim, M.-O. Changes in Surface Electromyography Signal According to Severity in Patients with Carpal Tunnel Syndrome. *J. Electrodiagn. Neuromuscul. Dis.* **2020**, *22*, 15–22. [CrossRef]
33. Chowdhury, R.; Reaz, M.B.I.I.; Islam, M.T. Wavelet Transform to Recognize Muscle Fatigue. In *Proceedings of the 2012 Third Asian Himalayas International Conference on Internet*; IEEE: Kathmundu, Nepal, 2012; pp. 1–5. [CrossRef]
34. Dantas, J.L.; Camata, T.V.; Brunetto, M.A.O.C.; Moraes, A.C.; Abrão, T.; Altimari, L.R. Fourier and Wavelet Spectral Analysis of EMG Signals in Isometric and Dynamic Maximal Effort Exercise. In *Proceedings of the 2010 Annual International Conference of the IEEE Engineering in Medicine and Biology Society, EMBC'10*; IEEE: Buenos Aires, Argentina, 2010; pp. 5979–5982. [CrossRef]
35. Yousif, H.A.; Zakaria, A.; Rahim, N.A.; Salleh, A.F.B.; Mahmood, M.; Alfarhan, K.A.; Kamarudin, L.M.; Mamduh, S.M.; Hasan, A.M.; Hussain, M.K. Assessment of Muscles Fatigue Based on Surface EMG Signals Using Machine Learning and Statistical Approaches: A Review. In *IOP Conference Series: Materials Science and Engineering*; IOP Publishing Ltd.: Bristol, UK, 2019; Volume 705.
36. Chang, K.M.; Liu, S.H.; Wu, X.H. A Wireless SEMG Recording System and Its Application to Muscle Fatigue Detection. *Sensors* **2012**, *12*, 489–499. [CrossRef]
37. Kuthe, C.D.; Uddanwadiker, R.V.; Ramteke, A.A. Surface Electromyography Based Method for Computing Muscle Strength and Fatigue of Biceps Brachii Muscle and Its Clinical Implementation. *Inform. Med. Unlocked* **2018**, *12*, 34–43. [CrossRef]
38. Siecinski, S.; Kostka, P.S.; Tkacz, E.J. Assessment of Muscle Fatigue During The Walk on a Treadmill in Healthy Subjects Based on Median Frequency of Surface Electromyogram. In *Proceedings of the 2022 IEEE 21st Mediterranean Electrotechnical Conference (MELECON)*; IEEE: Palermo, Italy, 2022; pp. 436–441. [CrossRef]
39. Macisaac, D.; Parker, P.A.; Scott, R.N. The Short-Time Fourier Transform and Muscle Fatigue Assessment in Dynamic Contractions. *J. Electromyogr. Kinesiol.* **2001**, *11*, 439–449. [CrossRef]
40. Sparto, P.J.; Parnianpour, M.; Barria, E.A.; Jagadeesh, J.M. Wavelet and Short-Time Fourier Transform Analysis of Electromyography for Detection of Back Muscle Fatigue. *IEEE Trans. Rehabil. Eng.* **2000**, *8*, 433–436. [CrossRef]
41. Boyer, M.; Bouyer, L.; Roy, J.S.; Campeau-Lecours, A. A Real-Time Algorithm to Estimate Shoulder Muscle Fatigue Based on Surface EMG Signal for Static and Dynamic Upper Limb Tasks. In *Proceedings of the Annual International Conference of the IEEE Engineering in Medicine and Biology Society, EMBS*; IEEE: Mexico City, Mexico, 2021; pp. 100–106. [CrossRef]
42. Thongpanja, S.; Phinyomark, A.; Phukpattaranont, P.; Limsakul, C. Mean and Median Frequency of EMG Signal to Determine Muscle Force Based on Time Dependent Power Spectrum. *Elektron. Elektrotechnika* **2013**, *19*, 51–56. [CrossRef]
43. Zhang, X.; Wang, Y.; Han, R.P.S. Wavelet Transform Theory and Its Application in EMG Signal Processing. In *Proceedings of the 2010 Seventh International Conference on Fuzzy Systems and Knowledge Discovery*; IEEE: Yantai, China, 2010; Volume 5, pp. 2234–2238. [CrossRef]
44. Chowdhury, S.K.; Nimbarte, A.D.; Jaridi, M.; Creese, R.C. Discrete Wavelet Transform Analysis of Surface Electromyography for the Fatigue Assessment of Neck and Shoulder Muscles. *J. Electromyogr. Kinesiol.* **2013**, *23*, 995–1003. [CrossRef] [PubMed]
45. Vannozzi, G.; Conforto, S.; D'Alessio, T. Automatic Detection of Surface EMG Activation Timing Using a Wavelet Transform Based Method. *J. Electromyogr. Kinesiol.* **2010**, *20*, 767–772. [CrossRef] [PubMed]
46. Karlsson, S.; Gerdle, B. Mean Frequency and Signal Amplitude of the Surface EMG of the Quadriceps Muscles Increase with Increasing Torque—A Study Using the Continuous Wavelet Transform. *J. Electromyogr. Kinesiol.* **2001**, *11*, 131–140. [CrossRef]

https://doi.org/10.3390/bioengineering13020248

47. Canal, M.R. Comparison of Wavelet and Short Time Fourier Transform Methods in the Analysis of EMG Signals. *J. Med. Syst.* **2010**, *34*, 91–94. [CrossRef]
48. Reaz, M.B.I.; Hussain, M.; Mohd-Yasin, F. EMG Analysis Using Wavelet Functions to Determine Muscle Contraction. In *Proceedings of the 8th International Conference on e-Health Networking, Applications and Services*; IEEE: New Delhi, India, 2006; pp. 132–134. [CrossRef]
49. Wang, J.; Sun, S.; Sun, Y. A Muscle Fatigue Classification Model Based on LSTM and Improved Wavelet Packet Threshold. *Sensors* **2021**, *21*, 6369. [CrossRef]
50. Li, Z.; Wang, X.; Li, Q.; Wang, F.; Tao, X. Muscle Fatigue Identification and Prediction in Motion Using Wearable Device with Power and Torque-Based Features. *Wearable Electron.* **2025**, *2*, 62–68. [CrossRef]
51. Murillo-Escobar, J.; Jaramillo-Munera, Y.E.; Orrego-Metaute, D.A.; Delgado-Trejos, E.; Cuesta-Frau, D. Muscle Fatigue Analysis during Dynamic Contractions Based on Biomechanical Features and Permutation Entropy. *Math. Biosci. Eng.* **2020**, *17*, 2592–2615. [CrossRef]
52. Al-Mulla, M.R.; Sepulveda, F.; Colley, M. A Review of Non-Invasive Techniques to Detect and Predict Localised Muscle Fatigue. *Sensors* **2011**, *11*, 3545–3594. [CrossRef]
53. Ghofrani Jahromi, M.; Parsaei, H.; Zamani, A.; Dehbozorgi, M. Comparative Analysis of Wavelet-Based Feature Extraction for Intramuscular EMG Signal Decomposition. *J. Biomed. Phys. Eng.* **2017**, *7*, 365–378. [PubMed]
54. Achmamad, A.; Jbari, A. A Comparative Study of Wavelet Families for Electromyography Signal Classification Based on Discrete Wavelet Transform. *Bull. Electr. Eng. Inform.* **2020**, *9*, 1420–1429. [CrossRef]
55. Berger, P.D.A.; Nascimento, F.A.D.O.; Do Carmo, J.C.; Da Rocha, A.F. Compression of EMG Signals with Wavelet Transform and Artificial Neural Networks. *Physiol. Meas.* **2006**, *27*, 457–465. [CrossRef]
56. Ozsert, M.; Yavuz, O.; Durak-Ata, L. Analysis and Classification of Compressed EMG Signals by Wavelet Transform via Alternative Neural Networks Algorithms. *Comput. Methods Biomech. Biomed. Eng.* **2011**, *14*, 521–525. [CrossRef]
57. Shilaskar, S.; Bhatlawande, S.; Chavare, R.; Ingale, A.; Joshi, R.; Vaishale, A. Human Hand Movement Classification Based on EMG Signal Using Different Feature Extractor. *Biomed. Pharmacol. J.* **2024**, *17*, 71–82. [CrossRef]
58. Rong, Y.; Hao, D.; Han, X.; Zhang, Y.; Zhang, J.; Zeng, Y. Classification of Surface EMGs Using Wavelet Packet Energy Analysis and a Genetic Algorithm-Based Support Vector Machine. *Neurophysiology* **2013**, *45*, 39–48. [CrossRef]
59. Aljebory, K.M.; Jwmah, Y.M.; Mohammed, T.S. Classification of EMG Signals: Using DWT Features and ANN Classifier. *IAENG Int. J. Comput. Sci.* **2024**, *51*, 23–31.
60. Côté-Allard, U.; Fall, C.L.; Drouin, A.; Campeau-Lecours, A.; Gosselin, C.; Glette, K.; Laviolette, F.; Gosselin, B. Deep Learning for Electromyographic Hand Gesture Signal Classification Using Transfer Learning. *IEEE Trans. Neural Syst. Rehabil. Eng.* **2019**, *27*, 760–771. [CrossRef] [PubMed]
61. Arozi, M.; Putri, F.T.; Ariyanto, M.; Caesarendra, W.; Widyotriatmo, A.; Setiawan, J.D. Electromyography (EMG) Signal Recognition Using Combined Discrete Wavelet Transform Based Adaptive Neuro-Fuzzy Inference Systems (ANFIS). In *AIP Conference Proceedings*; AIP Publishing LLC: Melville, NY, USA, 2017; Volume 1788. [CrossRef]
62. Tuncer, T.; Dogan, S.; Subasi, A. Surface EMG Signal Classification Using Ternary Pattern and Discrete Wavelet Transform Based Feature Extraction for Hand Movement Recognition. *Biomed. Signal Process. Control* **2020**, *58*, 101872. [CrossRef]
63. Aarotale, P.N.; Rattani, A. Time Frequency Analysis of EMG Signal for Gesture Recognition Using Fine Grained Features. *arXiv* **2025**, arXiv:2504.14708. [CrossRef]

Disclaimer/Publisher's Note: The statements, opinions and data contained in all publications are solely those of the individual author(s) and contributor(s) and not of MDPI and/or the editor(s). MDPI and/or the editor(s) disclaim responsibility for any injury to people or property resulting from any ideas, methods, instructions or products referred to in the content.

https://doi.org/10.3390/bioengineering13020248

bioengineering

MDPI

Article

The Effects of Neuromuscular Training on Electromyography, Lower Extremity Kinematics, and Ground Reaction Force During an Unanticipated Side-Cut on Recreational Female Hockey Players

Tom Johnston [1,*], Stephanie Valentin [2], Susan J. Brown [1] and Konstantinos Kaliarntas [1,3]

1 School of Applied Sciences, Sighthill Campus, Edinburgh Napier University, Edinburgh EH11 4BN, UK; su.brown@napier.ac.uk (S.J.B.); kaliarntas@upatras.gr (K.K.)
2 School of Health and Social Care, Sighthill Campus, Edinburgh Napier University, Edinburgh EH11 4BN, UK; s.valentin@napier.ac.uk
3 Department of Physiotherapy, University of Patras, GR26504 Patras, Greece
* Correspondence: t.johnston@napier.ac.uk

Abstract

During an unpredictable side-cut, this study examined how a sport-specific neuromuscular training program (NMTP) influenced electromyography responses in the lower limb posterior muscles, leg movement angles, maximum vertical ground reaction force (vGRF), and the rate of force development of vGRF. Thirty-eight adult female recreational hockey players were randomly allocated into an intervention group (INT) or a control group (CON). Before beginning training or matches, the INT carried out the NMTP three times per week for eight weeks, whereas the CON performed their routine warm-up. A 45° sidecut (dominant leg only) was performed at baseline and after eight-weeks and recorded with a motion capture system. The effect of group and time, and their interaction, was investigated using a mixed-design ANOVA. After landing, the participants in the INT had greater activation of their gastrocnemius lateralis, gastrocnemius medialis, and gluteus maximus muscles than those in the CON. INT participants showed significantly lower amounts of maximum knee abduction and knee excursion, while there was an increase in these variables for the CON. At week eight, the vGRF RFD decreased for the INT but increased for the CON. Although non-significant, the overall muscle activity showed an increasing trend for the INT when it came to supervised NMTP for eight weeks compared to the effect seen in the CON. This activity caused greater alterations in the motion and forces of the lower body for the INT than the CON.

Keywords: biomechanics; neuromuscular training; field hockey; injury prevention; side-cut

1. Introduction

Field hockey is a high-intensity, intermittent, invasive Olympic team sport [1]; therefore, there are a considerable number of injuries. The injury rates range from 0.1 to 90.9 per 1000 h played [2] and most injuries—up to 64%—that occur are 'non-contact' [2–4].

Most injuries in hockey occur to the lower extremity, most commonly to hamstring (0.8/1000 Athlete-Exposures) [5] and knees (0.57/Athlete-Exposures) and with ankle (0.76/1000 Athlete-Exposures) [6] and hip joints also frequently affected [2,4,7]. The injuries cause muscle strains and ligament damage [8,9]. The mechanisms involved in the most non-contact injuries are side-cutting, acceleration, deceleration and landing [9], similar to other team sports [10]. The proportion of non-contact and frequency of injuries is of concern, and therefore injury reduction mechanisms and programs require further attention [2].

Injury prevention programs have been shown to be useful in reducing injury rates in team sports [11–15] including hockey [16,17]. Injury prevention programs in hockey have reduced injuries (hazard ratio of 0.64) and the injury burden, a reduction of 8.42 days lost per 1000 player-hours [16] in youth field hockey players. Also, in elite female hockey players, reduced Anterior Cruciate Ligament (ACL) injury rates (0.4 per 1000 player hours in the control year) to zero in the intervention period was effective in reducing knee valgus moments and muscle activation [17].

Both the above studies used a combination of targeted training activities, known as neuromuscular training (NMT), which can improve certain characteristics (i.e., motor control, strength, etc) and potentially reduce the risk for various injuries [18]. Through its implementation, there has been a 26% decrease in all injuries, a 32% decrease in ankle sprains, a 61% decrease in ACL injuries, and reductions of up to 70% in hamstring injuries [19].

The most promising NMT programs appear to be multi-modal. Those programs, including plyometric training, improving body control (core stability), strength, balance, and sports specific movements, benefit athletes the most. The multi-modal programs that also have instructor feedback appear to produce even more effective results [20,21]. The optimal dose appears to be at least 20 min, twice a week, in order a NMT to elicit positive results [22], i.e., increasing muscle activation, which can also alter kinematics and landing forces.

Increasing muscle activation during landing through NMT can potentially reduce the injury risk [17], and Parsons [18] observed that gluteal muscle activation following NMT helped to reduce ACL injury risks by lowering hip adduction and reducing ground reaction forces when landing. A softer landing helps to reduce forces at the hip, rather than letting ligaments and similar structures do this work [23]. In field hockey, NMT programs have made positive impacts. Throughout a season, youth players who underwent an NMT intervention were less likely to suffer injuries [16]. In contrast, NMT with elite women resulted in reduced likelihood of both lower limb injuries and considerably lower knee valgus at its peak [17]. This was probably because NMT caused greater muscle activation in large muscle groups, such as the gluteal muscles, resulting in greater stability of the lower limb joints during activities such as landing.

Although there are many studies assessing the effects of NMT on females and two studies assessing the effects of NMT on hockey players, there are, to the authors knowledge, no studies assessing how NMT affects the biomechanics of recreational female hockey players during an unanticipated cut. Therefore, the aim of the current study is to investigate how an eight-week NMT program affect EMG, kinematics, and kinetics of recreational female hockey players during an unexpected side-cut. We hypothesized that implementing a NMT would result in greater muscle activity during take-off and landing (primary result), improved injury-related movement patterns (secondary result), and less force (tertiary result) during landing compared to the control group performing their usual exercises.

2. Materials and Methods

2.1. Study Design and Participants

A controlled-trial design used in this study. Experienced female university athletes who play hockey were recruited (via posters, hockey websites, and word of mouth), and those on the same team were divided into the control and intervention groups. Hockey experience, participating in training and games at least 3 times per week for at least a year, not having any injuries for 3 months, and agreeing to be part of the study were all included as eligibility requirements. If people had any current or past musculoskeletal injuries that impacted their movements, they were excluded from the study.

This study was approved under the ethical standards of Edinburgh Napier University and the School of Applied Sciences. Donnelly et al. [24] demonstrated that for the current dataset (using G*Power®, Los Angeles, CA, USA version 3.1.9.3), 14 subjects per group were required to provide 80% power at a 5% alpha level.

2.2. Intervention and Implementation

The intervention involved an evidence-based NMT program 3 times weekly for 8 weeks, presented following the TIDieR guidelines [25] (Supplementary Material https://www.mdpi.com/article/doi/s1). It was informed by previously published works [14,22,26–31]. The NMTP taught the players new ways to run and land, engage muscles for better mobility, strengthen their core and lower body, improve balance, and work on agility and sport specific movements. All the elements appeared in situations that were both expected and unexpected. The only equipment used was everyday hockey items: hockey sticks, balls, and cones.

The primary researcher (TJ) carried out the intervention to ensure consistency. The players received feedback, which was given using external coaching methods, to help improve their skills [32]. At the beginning of each training session and match, athletes in INT completed the Neuromuscular Training Program 3 days each week instead of performing their typical warm-up [33]. For each training session, the players were advised to complete all the exercises so that, step by step, their quality would increase and each activity would be performed vigorously (that is, rated 8 out of 10 according to Borg's scale [34]). This was possible when the participants were guided, motivated, and given feedback by a trained instructor during the intervention [21]. During the program, the number of completed sessions was regularly recorded.

The intervention consisted of a (1) pulse-raising section with (2) muscle activation into a (3) mobilization section. The program then has a (4) core stability phase, then a (5) balance phase, followed by a (6) plyometrics and strength phase, then on to (7) agility and (8) potentiation phases. The warm-up concluded with a sport-specific movement section. The total time for the program was 20 min (Supplementary Information).

For the warm-up during the intervention, the control group carried on with 2 min of pulse raising, different static and dynamic stretches, and hockey-related skills and moves related to their teams' usual activities, unlike the intervention group's NMTP.

2.3. Test Protocol

All participants performed an unanticipated cut (USC) at the baseline and 8 weeks later. Every testing session began with screening (PARQ) and height and body mass measurements, followed by performing a standard warm-up and practicing the sidecut. Before week 8, every player completed a questionnaire on the amount of hockey played, physical training and record any injuries. Testing for the study took place in the biomechanics laboratory at the Sighthill Campus of Edinburgh Napier University.

2.4. Unanticipated Side-Cut (USC)

The USC is a frequent sports activity, and its reliability has been tested before since it is also a frequent cause of injuries [35]. The direction of travel was determined by the timing gates positioned before the force plate and all participants performed five fast 45° side-cuts towards the indicated side (indicated by SmartSpeed Pro™, Fusion Sport®, Highlands Ranch, CO, USA) at the highest possible speed (minimal speed included was 2.14 m/s) (Figure 1a,b).

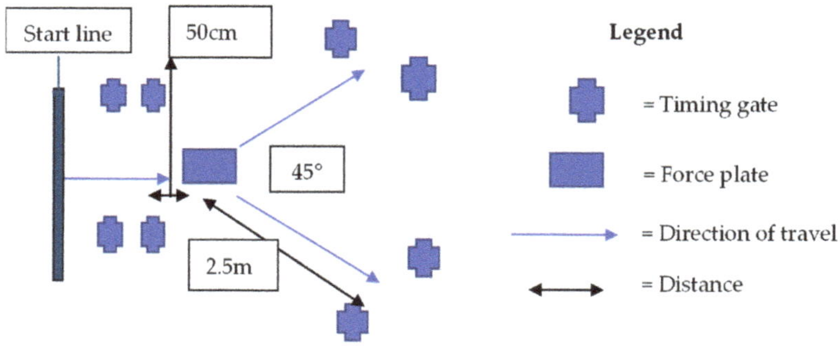

(a)

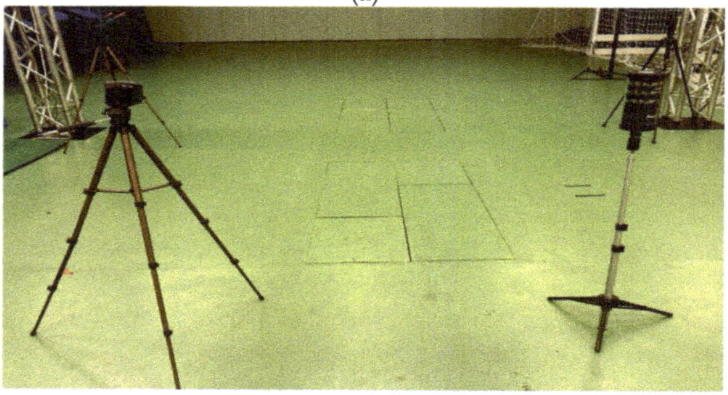

(b)

Figure 1. The set-up for the unanticipated side-cut task (**a**). The set-up of the equipment in the laboratory (**b**).

3. Data Collection and Processing

3.1. EMG

For this study, two Delsys Trigno Wireless™ sensors (SP-W01D, Natick, MA, USA), set to 1925.925 Hz (up-sampled to 2000 Hz), were applied to Gluteus Maximus (GMax), Gluteus Medius (GMed), Biceps Femoris (BF), Semitendinosus (ST), Gastrocnemius (medial) (GasMed), and Gastrocnemius (lateral) (GasLat), since the skin was prepared after shaving, alcohol cleaning, and drying. All sensors were installed according to the requirements outlined in SEMIAN guidelines [36]. Root Mean Square (RMS) data, which were acquired using a 30 ms window, were computed from the EMG signal after it had been transferred to EMGWorks Analysis™, Delsys, MA, USA. The EMG data for every muscle was normalized with the highest value found among the five USC tasks [37,38].

3.2. Kinematic Data

Sixty-four 19 mm single and cluster reflective markers were positioned to anatomic landmarks and segments for kinematic measurements (Table 1 and Figure 2).

Twelve Oqus 300 motion capture cameras recorded three-dimensional data at 500 frames per second through the Qualysis Track Manager™ (QTM, Goteborg, Sweden). The data was then exported to the Visual 3D v6 software for processing (C-motion™, Germantown, MD, USA) program. Winter [39] and Roewer et al. [40] showed that differences in movement artifact is reduced with a filter of 10 Hz for all the leg segments, so a 10 Hz Butterworth bi-directional (4th order) filter was applied to the kinematic data. For the analysis in the present paper, lower body data (hip and knee) were used in the sagittal and frontal planes at the beginning of contact and just before maximum knee flexion.

Table 1. Reflective marker placement.

Anatomical Site	Marker Placement (No. of Markers)
Head	Left and Right Anterior and Posterior Cranium (4)
Trunk (Treated as a Single Segment)	Acromion Process (2) Posterior Superior Iliac Spine—PSIS (2) Upper Thoracic Cluster (4)
Arm and Finger (Bilateral)	Acromion Process (1) Lateral and Medial Elbow Epicondyle (2) Lateral and Medial Aspects of the Styloid Process (2) Third Metacarpal (Distal End) (1)
Pelvis (a CODA Pelvis)	Anterior Superior Iliac Spine—ASIS (2) Posterior Superior Iliac Spine—PSIS (2) Sacrum
Thigh (Bilateral)	Greater Trochanter (1) Lateral and Medial Femoral Condyle (2) Thigh Cluster (4)
Shank (Bilateral)	Lateral and Medial Femoral Condyle (2) Lateral and Medial Malleolus (2) Shank Cluster (4)
Foot (Bilateral)	Lateral and Medial Malleolus (2) Heel (1) 1st and 5th Metatarsal (2)

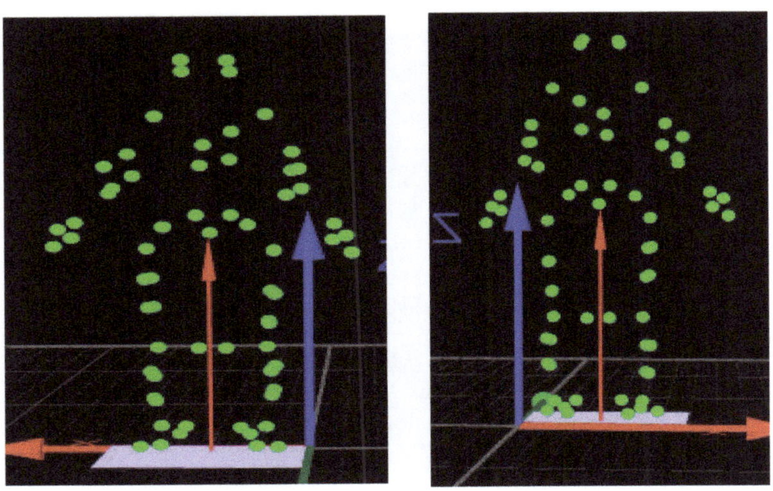

(a) (b)

Figure 2. Green dots represent the reflective marker placement—(**a**) anterior view and (**b**) posterior view (depicted by Qualisys). Arrows show the global laboratory Axis reference system and the resultant ground reaction force.

3.3. Kinetics

A Kistler force plate (Kistler Instruments Ltd.®, Model 569B, Winterthur, Switzerland) was used for ground reaction forces measurement, and the data were sampled at 1000 Hz. The 20 Hz Butterworth bi-directional (4th order) filter was applied to the movement data.

All data was synchronized in QTM. Comparable to numerous studies, this filter level falls within the typical range of 12 Hz and 50 Hz [27,41].

3.4. Statistical Analysis

Data presented descriptively using means and standard deviations. The data was investigated using diagnostic statistics such at boxplots for outliers and Shapiro-wilk tests for normality. An independent samples *t*-test was employed to explore differences among group baseline characteristics. An ANOVA with three factors was performed in SPSS® for the EMG data of every muscle: (1) a between-subjects factor: group; (2) a within-subjects factor: time before and after the landing; and (3) a within-subjects factor: the landing phase split between 30 ms before landing and 50 ms afterwards. If any main effects or interactions were significant, pairwise comparisons with Bonferroni corrections were carried out to identify differences. For variables such as maximum knee abduction, knee excursion, vGRF, normalized vGRF, and RFD, Two-Way mixed ANOVAs were employed by considering group as the between-subjects factor and time as the within-subjects factor. Analysis results were significant when the α-level was <5% (two-tailed). The authors mentioned the effect sizes (partial eta squared, ηp^2) and pairwise test comparisons with 95% confidence intervals when appropriate.

4. Results

Between September 2016 and April 2017, a total of 51 people were recruited, and after the loss of 13 participants along the way (Figure 3), the final 38 participants (INT = 18, CON = 20) completed the study (Table 2). Most measured variables did not differ significantly when we compared the groups (Table 2); however, there was a significant difference in experience between the groups (t37, −2.2, $p = 0.034$, confidence interval (CI) = −3.72 to −0.155). There were no reported adverse events or injuries, and the overall compliance to the NMT program was 66.9%.

Table 2. Participant characteristics (Mean ± SD).

Variables	Groups		
	CON	INT	*p*-Value
Age (yrs)	20.04 (1.6)	19.9 (1.1)	0.97
Height (cm)	165.2 (4.7)	167.6 (5.4)	0.40
Body mass (kg) (baseline)	62.9 (7.8)	66.0 (6.3)	0.68
Body mass (kg) (8 weeks)	62.7 (7.8)	66.4 (6.4)	0.74
No. of games per week (n)	1.5 (0.6)	1.8 (0.4)	0.09
No. of training sessions (hockey) (n)	1.21 (0.4)	1.0 (0.3)	0.92
No. of training sessions per week (not hockey) (n)	1.0 (0.6)	1.1 (0.8)	0.43
Playing experience (years)	9.3 (3.1)	11.2 (2.4)	0.03
No. of injuries in intervention period (n)	0.2 (0.4)	0.1 (0.3)	0.32

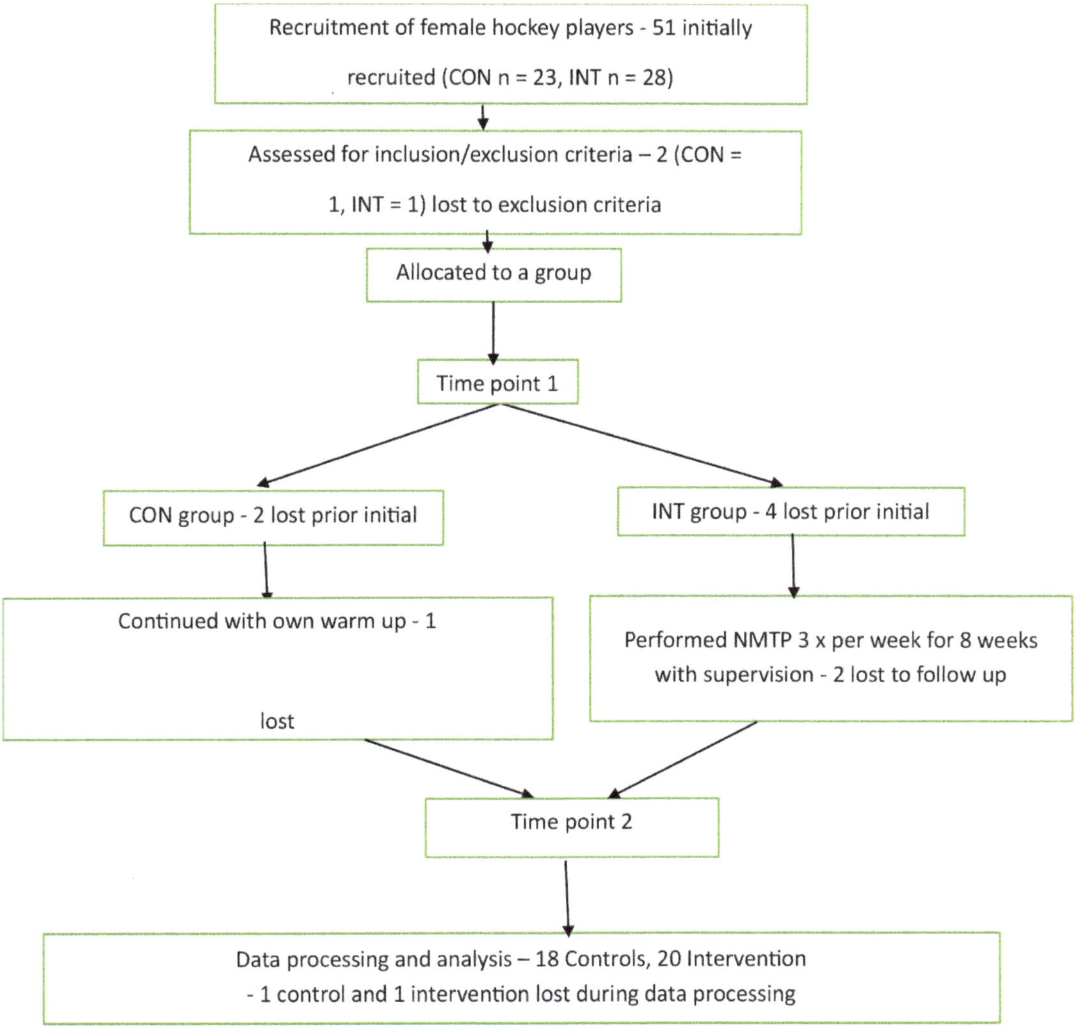

Figure 3. Flow diagram of participants through the study (CON—control; INT—intervention; NMTP—neuromuscular training program).

4.1. EMG

GasMed: There was no significant difference between the groups at 30 ms before IC (F (1,36) = 0.127, p = 0.723, $\eta p^2 = 0.004$). After 50 ms of IC, the amount of muscle activation increased in both groups (CON = 19.9, SD = 11.7 to 20.2, SD 10.4; INT = 25, SD 11.1 to 28.8, SD 10.8) and although non-significant it was higher in the INT (F (1,36) = 1.037, p = 0.315, $\eta p^2 = 0.028$) (Table 3).

GasLat: There was no significant difference 30 ms prior to landing (F (1,36) = 1.159, p = 0.298, $\eta p^2 = 0.031$). After 50 ms from the initial contact, there was a noticeable difference between the groups. Both groups showed more muscle activity (CON = 18.7% (SD 10.1) to 20.8% (SD 10.5); INT = 24.1% (SD 12.4) to 28.3% (SD 8.1), with a greater increase of 4.2% in the INT than the CON, rising by just 2.1% (F (1,36) = 1.844, p = 0.183, $\eta p^2 = 0.490$).

Table 3. Normalized EMG during the unanticipated side-cut.

Muscle	Time Point	Group	Normalized EMG (%) Mean (SD)		*p*-Value
			Baseline	Week 8	Interaction
GasMed	30 ms prior to IC	CON INT	25.3 (14.9) 18.3 (11.6)	28.4 (11.9) 23.1 (16.7)	0.723
	50 ms after IC	CON INT	19.9 (11.7) 25.1 (11.1)	20.2 (10.4) 28.8 (10.8)	0.315
GasLat	30 ms before IC	CON INT	27.1 (15.2) 17.9 (9.9)	25.9 (9.9) 24.0 (15.0)	0.298
	50 ms after IC	CON INT	18.7 (10.1) 24.1 (12.4)	20.8 (10.5) 28.3 (8.1)	0.183
ST	30 ms before IC	CON INT	35.9 (11.5) 35.7 (15.1)	31.1 (9.9) 34.2 (18.6)	0.291
	IC to 50 ms	CON INT	37.2 (11.6) 37.8 (11.5)	31.1 (8.2) 34.5 (9.8)	0.030
BF	30 ms before IC	CON INT	36.3 (13.7) 33.7 (15.2)	37.1 (15.2) 34.4 (11.4)	0.800
	IC to 50 ms	CON INT	35.6 (11.1) 35.5 (12.2)	33.4 (9.9) 35.6 (9.6)	0.567
GMed	30 ms before IC	CON INT	27.9 (11.7) 24.7 (7.1)	21.9 (6.1) 22.1 (8.4)	**0.013**
	IC to 50 ms	CON INT	29.2 (10.8) 23.2 (8.2)	24.5 (6.1) 23.9 (4.7)	**0.035**
GMax	30 ms before IC	CON INT	29.2 (12.6) 28.9 (9.8)	24.4 (9.6) 24.9 (12.0)	0.067
	IC to 50 ms	CON INT	19.9 (11.7) 27.2 (11.9)	20.5 (10.1) 29.0 (10.6)	0.542

(GasMed = Gastrocnemius (medial), GasLat = Gastrocnemius (Lateral), ST = Semitendinousus, BF = Biceps Femoris, IC = Initial Contact, millisseconds, CON = control group, INT = intervention group).

ST: There was no difference, whether interaction or main effects, in these muscles at 30 ms just before landing (F (1,36) = 1.149, p = 0.291, ηp^2 = 0.031). There was a significant difference at 50 ms after landing (F (1,36) = 5.116, p = 0.030, ηp^2 = 0.031) with both groups decreasing the muscle activation. The INT had a smaller reduction in EMG (CON = 6.2, INT = 3.3).

BF: Results from a mixed-design ANOVA demonstrated that there was no significant difference between the two groups at either point in time (F1,36) = 0.065, p = 0.800, ηp^2 = 0.002, F (1,36) = 0.333, p = 0.567, ηp^2 = 0.008 respectively). Both groups experienced approximately the same increase in muscle activation for both time points: CON from 30 ms before IC—from 29.1% (10.2) to 30.5% (17.0); INT pre-test from 32.3% (11.1) to 33.05% (11.3); CON from IC to 50 ms after—from 35.6% (11.1) at pre-test to 33.4% (9.9); INT from IC to 50ms after = 35.5% (12.2) at pre-test to 35.6 (9.6) at post-test.

GMed: There was a significant interaction effect for 30 ms before landing (F (1,36) = 6.759, p = 0.013, ηp^2 = 0.290) and 50 ms after landing (F (1,36) = 4.810, p = 0.035, ηp^2 = 0.118). Between the CON and INT, there was less reduction in muscle activation 30 ms before landing in the INT compared to the CON (2.6% vs. 6%), which was statistically significant (p = 0.013, F (1,36) = 6.759, ηp^2 = 0.158). Following the landing, there was a drop of 4.7% in the CON but a similar or slight rise (0.7%) for the INT (F (1,36) = 4.810, p = 0.035, ηp^2 = 0.118).

GMax: There was no significant difference at 30 ms before landing for this muscle, as both groups showed decreases in GMax, with the INT having a smaller reduction (~4%), less than the CON, which lowered it by nearly 5% (F1,36) = 3.577, p = 0.067, ηp^2 = 0.090). There was no significant difference in post-landing electromyography at IC. (p = 0.542). Both groups saw an increase in muscle activation, with the INT increasing more, from 43.1% to 45.3% (SD 10.0 to 18.9), compared to the CON's increase from 19.9% to 20.5% (SD 11.7 to 10.1) (F1,36) = 0.378, p = 0.542, ηp^2 = 0.010).

4.2. EMG Results, Kinematics Results, Kinetics Results

No differences found in the sagittal or frontal planes of hip and knee measurements were found at initial or maximum knee flexion, when comparing groups or looking within groups, including maximum knee abduction (Tables 4 and 5). Specifically, no significant differences were found in hip flexion (F (1,36) = 3.713, p = 0.062, ηp^2 = 0.093;) or knee flexion (F (1,36) = 0.085, p = 0.772, ηp^2 = 0.002) at IC. In addition, at IC in the frontal plane, the hip (Lateral Flexion—F (1,36) = 0.533, p = 0.47, ηp^2 = 0.015 F (1,36) = 2.685, p = 0.110, ηp^2 = 0.099) and the knee abduction (F (1,36) = 1.466, p = 0.234, ηp^2 = 0.039) did not show any significant differences.

Table 4. Lower extremity sagittal and frontal plane kinematics at initial contact (IC).

Joint	Time Point	Variable	Group	Value		
				Before	Week 8	Interaction
Hip	IC	Flexion	CON INT	38.86 (8.6) 41.99 (9.8)	37.85 (8.7) 35.77 (8.7)	0.853
	IC	Lateral flexion	CON INT	12.74 (5.4) 10.87 (6.4)	10.87 (6.4) 9.72 (5.4)	0.291
Knee	IC	Flexion	CON INT	20.70 (6.7) 19.28 (10.1)	18.64 (8.4) 17.96 (8.2)	0.193
Knee	IC	Abduction	CON INT	0.27 (4.5) 1.01 (4.3)	0.77 (4.9) 1.9 (3.1)	0.732

Table 5. Lower extremity sagittal and frontal plane kinematics at maximum knee flexion (MKF).

Joint	Time Point	Variable	Group	Value		
				Before	Week 8	Interaction
Hip	MKF	Flexion	CON INT	45.7 (11.7) 45.2 (11.1)	46.1 (11.0) 38.8 (8.0)	0.073
	MKF	Lateral flexion	CON INT	13.02 (7.3) 11.4 (6.6)	12.4 (7.2) 10.4 (6.2)	0.370
Knee	MKF	Flexion	CON INT	61.8 (7.2) 62.0 (11.5)	63.6 (6.9) 63.6 (6.8)	0.680
	MKF	Abduction	CON INT	−1.09 (6.7) −3.81 (7.6)	−3.65(7.1) −5.68 (6.6)	0.055

No significant differences were found for hip and knee flexion values at MKF (hip: F (1,36) = 3.401, p = 0.073, ηp^2 = 0.086; knee: F (1,36) = 0.172, p = 0.680, ηp^2 = 0.005). Similarly, the abduction/adduction movement at the hip (F (1,36) = 0.835, p = 0.367, ηp^2 = 0.023) and the knee (F (1,36) = 4.308, p = 0.055, ηp^2 = 0.107) did not significantly differ.

However, regardless of the lack of statistically significant differences, there were some trends worth noting in maximum knee abduction and knee excursion after the

8 weeks (Figure 4). There was a significant within-subjects main effect for maximum knee abduction (F (1,36) = 7.721, $p = 0.009$, $\eta p^2 = 0.18$), as well as a substantial time-by-group interaction (F (1,36) = 8.096, $p = 0.007$, $\eta p^2 = 0.18$). The INT group experienced a decrease in maximum knee abduction after eight weeks of Neuromuscular Training (NMT) (Week 8: INT Mean = 8.47, SD = 2.8) whereas the control group experienced an increase.

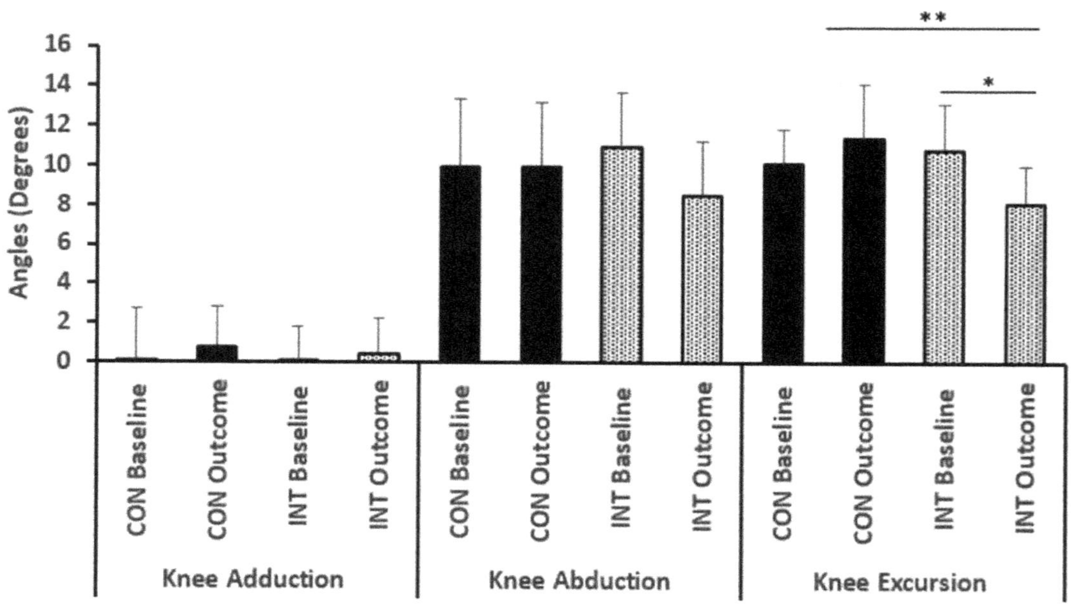

Figure 4. Maximum knee adduction, abduction, and excursion angles during an unanticipated side-cut (means ± SD). ** Significant interaction effect, * Significant within subjects main effect. INT Intervention, CON: control.

4.3. Kinetics

Peak vGRF and normalized vGRF showed no significant difference between the groups (F (1,36) = 0.042, $p = 0.838$, $\eta p^2 = 0.001$; F (1,36) = 0.408, $p = 0.527$, $\eta p^2 = 0.011$) (Table 6). The vGRF peak increased by more than 40 N in the CON group, and it decreased by about 55 N in the INT group.

Table 6. Peak vertical ground reaction force (vGRF) and normalized (to body weight) vertical ground reaction force (Norm vGRF).

Variable	Group	Value Mean (SD)		*p*-Value		
		Before	Week 8	Interaction	Within	Between
Peak vGRF (N)	CON INT	1185.2 (212.6) 1226.7 (208.6)	1226.3 (211.2) 1171.5 (180.5)	0.838	0.170	0.910
Norm vGRF (N/BW)	CON INT	1.97 (0.3) 1.90 (0.3)	1.98 (0.3) 1.82 (0.2)	0.527	0.360	0.100

RFD: The mean RFD was similar between the groups at pre-test (CON = 19.5 (9.9) vs. INT = 18.9 (9.1) BW/s). However, there was a significant time by group interaction; the mean RFD for the CON increased (by 2.22 BW's/s), whereas the mean for the INT decreased (by 6.08 BW's/s). At week 8: CON = 21.75 BWs/s, SD = 9.4; INT = 12.83 BWs/S, SD = 5.7) (within subjects main effect: F (1,36) = 2.476, $p = 0.124$, $\eta p^2 = 0.064$;

between groups effect: F (1,36) = 3.593, $p = 0.066$, $\eta p^2 = 0.091$; time by group interaction: (F (1,36) = 11.519, $p = 0.002$, $\eta p^2 = 0.242$) (Figure 5).

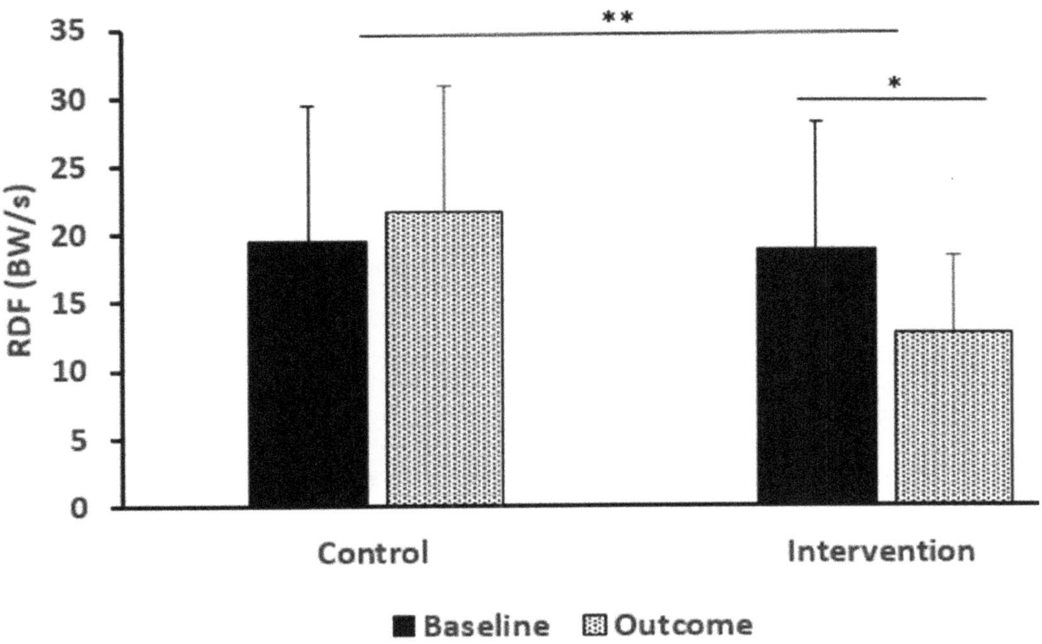

Figure 5. Rate of Force Development (RDF) during an unanticipated side-cut (body weight per second, BW/s) (means ± SD). ** Significant interaction effect, * Significant within subjects main effect. INT Intervention, CON: control.

5. Discussion

The study sought to explore how a NMTP could improve the movement skills of female hockey players. According to the hypothesis, neuromuscular training could potentially increase the activation profile in lower limb muscles, which could lead into more controlled movements and subsequently the players to be subjected in lower risk of non-contact injuries.

5.1. EMG

After the 8 weeks period, the normalized gluteus maximus (GMax) EMG was largely unchanged in both groups despite the inclusion of squats, lunges, and glute bridge exercises in the intervention, all of which have been reported to elicit GMax activation [42]. The gluteus medius (GMed) activation decreased in the CON, whereas the INT maintained the muscle activation. Although the intervention included hip abduction movements shown to produce gluteus medius muscle activation by DiStefano et al. [16] and Boren et al. [43], future interventions may benefit from more specific exercises to increase activation and therefore reduce hip adduction during landing in high intensity sports activities. These findings differ from Weir et al. [17] who reported an increase of 10% total gluteal muscle activation. This might be due to the population differences between the two studies (i.e., elite vs. recreational hockey players), the additional strength and conditioning sessions the elite players usually complete, the length of the intervention or a combination of these factors. In addition, the program followed by Weir et al. might have featured a greater proportion of glute specific exercises and measured total gluteal activity whereas this study included workouts for the whole lower limb.

The gluteal muscles are responsible for key movements such as hip extension, abduction and external rotation and hence for opposing adduction and internal rotation which are usually involved to common injury mechanisms [35,44,45]. The NMTP in this study included single-leg squats, arabesques, lunges, and plyometrics to assist this group of muscles. However, no significant increase in GMed activation was found after the intervention, which could suggest that more frontal plane specific exercises should be considered in future NMTP or the overall training volume and intensity of the programme needs further consideration.

The changes in hamstring muscle activation were not statistically significant when comparing baseline and outcome timepoints, even with the inclusion of Nordic hamstring curls and other eccentric contractions during the latter stages of the swing phase of sprinting and arabesque, all of which are high hamstring muscle activation exercises [46] and reduce injury rates [47]. A study by Zebis et al. [48] revealed that athletes involved in team sports experience increased ST activation after NMTP. Zebis et al. found there was no difference between INT and CON at pre-test in Semitendinosus activation however, there was a significant difference at post-test as the activation in the CON decreased whereas the INT increased the activation of this muscle. These results are different from our study possibly because the program duration (12 weeks vs 8 weeks), and maybe due to the lack of progression in this study. In addition, Zebis et al. [49] found that prompting the ST system before IC with a lead time of 30 ms is essential for using safe movement techniques to protect the ACL., Further investigation may be warranted to find the optimum training volume, training progression, for increased muscle hamstring activation.

There was an increase in the GasMed and GasLat activity between the groups from IC to 50 ms post-landing at week 8, with a greater increase in the INT (a significant difference at week 8). As the gastrocnemius helps control, along with other lower extremity muscles, both the knee and ankle [50,51], this increase may be important. This change may be due to the muscle activation exercises at the start of the NMTP, agility and plyometric exercises directly, as well as the calf raises, lunges, running, and sport-specific movement along with the advised change in technique. These exercises can improve both activation and gastrocnemius reactivity [52], as there is considerable loading and unloading, which may not have occurred in the control group. The increase in gastrocnemius activity, especially GasMed [53] with associated knee and ankle control, could help reduce the risk of injuries, as both joints are frequently injured in hockey [2,4].

Brunner et al. [54] and Zebis et al. [48] propose NMTP with a frequency of at least twice a week for six weeks. The researchers suggest that, for this type of training to be effective an increase in the dosage to four times per week is required. Furthermore, future studies may benefit from monitoring intensity, effort and commitment to the intervention similar to Weir et al. [17]. in this study may have been influenced by the degree of attention the participants gave and the effort they put into their workouts. Using EMG alone is not the only way to find out if muscles are changing for the better. Also, the intervention could be developed to have a hamstring focus such as more Nordic Hamstring Exercises, which have been shown to strengthen and stimulate hypertrophy [55].

5.2. Kinematics

There were no significant differences in the two groups' sagittal and frontal plane movements of the hip. Decreased hip abductor strength such as the gluteal muscles, can cause excessive hip movement. [56,57]. It has been reported that the hip mobility may be altered after NMT, specifically a 3-month strength-based NMTP can decrease the hip adduction and knee abduction during the single-leg triple hop (Baldon et al. [58]. Evidence indicates that hip adduction is related to knee abduction during landing [27,59]. The length of the intervention and nature of the task may limit the change in frontal plane motion.

The INT and CON had comparable movements in the sagittal plane at MKF and IC, respectively. This outcome may come from the participants receiving additional feedback from outside sources as part of the intervention. Using feedback through communication appears to be an optimal way to develop and use skills in playing sports [32]. Moreover, when augmented feedback is used in training, there has been a reduction in knee valgus in the frontal plane to reduce more compared to no feedback (37.23% vs. 26.7%) [60] and this is important since knee abduction significantly raises the risk for ACL injury [26,60,61]. Recent studies indicate that nearly all ACL injuries in football players were caused by movements in the knee with valgus loading. During tasks with no contact, valgus loading occurred at the knee for 67% of kicks and 50% of jumps [62]. Hopper et al. [63] indicate that after neuromuscular training, female netball players increased knee flexion at both IC and MKF, unlike the control group, who had decreased knee flexion. This may be due to the focus (all lower body exercises) of the NMTP in their study, additionally, there was feedback on the players technique.

5.3. Kinetics

A trend of reduced vGRF following NMT in the INT, with a slight increase in the CON during the USC was observed in this study which could imply that CON are at greater injury risk [63,64]. In addition, a strong relationship exists between vGRF and the knee abduction motion since it has been reported to be 20% higher in female athletes with ACL injuries [27]. The absolute vGRF for the CON in this study was similar to that of the injured cohort in Hewett et al.'s [27] study, however it fell far below the ACL's tension limit [65]. In addition, a decrease in vGRF (with greater knee flexion) can reduce the knee flexion–extension moment and therefore anterior tibial shear [66]. This may be due to the exercises and partly because of the augmented feedback the athletes received throughout the study since it has need reported that the use of feedback can enhance motor learning and reduce the risk of injury [32] Therefore, the trends in this study suggest that the holistic NMTP which was implemented for an optimal period can potentially enhance motor learning, optimize loading profiles in activities which involve landing and change of direction and could potentially reduce the risk of injury in athletes.

The RFD was significantly lower in the INT vs. CON (21.8 vs. 12.8 BWs/s at post-test). This is in line with another study using a NMTP [67] and might be an indication that active structures (i.e., muscles) can potentially absorb more energy than the passive structures (i.e., ligaments and joints). Therefore, switching the traditional warm-ups (usually involving sets of 2 min pulse raisers, stretching, low intensity hockey skills, and some specific team-play maneuvers) to a NMTP could possibly lead to valuable outcomes. It should be underlined that the landing strategy is very important in reducing the RDF and a heel-first landing strategy can significantly increase (~33%) the loading and by [68]. In this study, particular attention was given to providing feedback and coaching to the athletes to improve their landing technique as suggested by Benjaminse et al. [32].

5.4. Limitations and Future Studies

Since the study was not a randomized controlled trial (RCT), there is a possibility of bias and a type 2 error. Future studies could be conducted as an RCT to enhance internal validity. Although a power calculation was conducted in this study, a larger sample size could potentially improve the power of the study to detect existing differences. The attendance was monitored in this study the intensity was not measured (e.g., measuring heart rate or rate of perceived exertion). This measure could be included in future studies to monitor each session of the intervention, since it is well known that the training intensity is an important factor in training programs. The dosage effect by increasing the amount of time and effort a person devotes to the program might help their EMG, kinematics, and kinetics improve. In addition, the participants in this study were recreational players compared to trained elite athletes that are participating in other studies which makes comparing the datasets difficult.

Future studies could include a more targeted approach on gluteal and hamstring muscles, as there were limited activation changes. In addition, future studies may benefit from progressive overload throughout the intervention. Furthermore, this study focused on recreational hockey players and therefore, future studies could include elite hockey players or further investigation of the intervention with other team sport players.

6. Summary

Following 8 weeks of NMTP, in this study trends of higher muscle activation were observed especially for the gluteal and gastrocnemius muscles. Moreover, the findings include a reduction of the knee abduction during landing and the overall knee excursion during the USC. Finally, a more marked reduction of the vertical ground reaction force and rate of force development in the NMTP group as compared to the control group was also observed. Altogether, these findings might suggest improved motor control and hence possibly lower risk of sports related injury after the implementation of a NMTP in recreational hockey players. For this reason, further research in the design and implementation of evidence based NMTP in team sports is warranted.

Supplementary Materials: The following supporting information, which includes the intervention, can be downloaded at: https://www.mdpi.com/article/10.3390/bioengineering12101101/s1.

Author Contributions: All stages of planning and conducting the study, from design to implementation, were led by T.J., K.K., and S.J.B.; T.J. did the research, designed and developed the NMTP software, managed data collection and handling, and wrote the first version of the manuscript as part of their PhD project. They assisted T.J. with formatting files in QTM and Visual 3D; K.K., S.J.B., and S.V. shared their interpretations of the main points. Everyone on the team assisted with this document and provided critical comments for the last version. All authors have read and agreed to the published version of the manuscript.

Funding: T.J. was a self-funded PhD candidate.

Institutional Review Board Statement: The study was conducted in accordance with the Declaration of Helsinki, and approved by the Institutional Ethics Committee of Edinburgh Napier University (approval code—4680432 on 25/01/2016) for studies involving humans.

Informed Consent Statement: Informed consent was obtained from all subjects involved in the study.

Data Availability Statement: The data presented in this study are available on request from the corresponding author. The data are not publicly available due to ethical considerations.

Acknowledgments: The authors are grateful to everyone from Edinburgh Napier University, Edinburgh University, and Heriot-Watt University Hockey Clubs who participated in this research.

Conflicts of Interest: The authors declare no conflicts of interest.

References

1. Hiepen, L.; Bosserhoff, N.; Schaudig, F.; Heitzer, F.; Jäger, M.; Mayer, C. Functional Knee Stability in Elite Field Hockey Depends on Playing Class and Gender. *Sports Med. Int. Open* **2024**, *9*, a24172488. [CrossRef]
2. Barboza, S.D.; Nauta, J.; van der Pols, M.J.; van Mechelen, W.; Verhagen, E.A.L.M. Injuries in Dutch elite field hockey players: A prospective cohort study. *Scand. J. Med. Sci. Sports* **2018**, *28*, 1708–1714. [CrossRef] [PubMed]

3. Hollander, K.; Wellmann, K.; Eulenburg, C.; Braumann, K.; Junge, A.; Zech, A.; Schiller, F. Epidemiology of injuries in outdoor and indoor hockey players over one season: A prospective cohort study. *Br. J. Sports Med.* **2018**, *52*, 1091–1096. [CrossRef] [PubMed]
4. Rees, H.; McCarthy Persson, U.; Delahunt, E.; Boreham, C.; Blake, C. Epidemiology of injuries in senior men's field hockey: A two-season prospective observational injury surveillance study. *J. Sports Sci.* **2020**, *38*, 2842–2849. [CrossRef]
5. Silvers-Granelli, H.J.; Cohen, M.; Espregueira-Mendes, J.; Mandelbaum, B. Hamstring muscle injury in the athlete: State of the art. *J. ISAKOS* **2021**, *6*, 170–181. [CrossRef]
6. Dick, R.; Hertel, J.; Agel, M.A.; Grossman, J.; Marshall, S.W. Descriptive Epidemiology of Collegiate Men's Basketball Injuries, National Collegiate Athletic Association Injury Surveillance System, 1988–1989 Through 2003–2004. *J. Athl. Train.* **2010**, *42*, 194–201.
7. Agel, J.; Olson, D.E.; Dick, R.; Arendt, E.A.; Marshall, S.W.; Sikka, R.S. Descriptive Epidemiology of Collegiate Women's Basketball Injuries, National Collegiate Athletic Association Injury Surveillance System, 1988–1989 Through 2003–2004. *J. Athl. Train.* **2007**, *42*, 202–210.
8. Beynnon, B.D.; Vacek, P.M.; Murphy, D.; Alosa, D.; Paller, D. First-time inversion Ankle Ligament Trauma. The effects of Sex, Level of Competition, and Sport on the incidence of Injury. *Am. J. Sports Med.* **2001**, *33*, 1485–1491. [CrossRef]
9. Johnston, T.; Kaliarntas, K.; Brown, S.J.; Taylor, C. Noncontact Injuries, Rates, Mechanism and Occurrence in Field Hockey in Scotland. *Euro. J. Sports Exerc. Sci.* **2023**, *11*, 30–37.
10. Ekstrand, J.; Hägglund, M.; Waldén, M. Injury incidence and injury patterns in professional football: The UEFA injury study. *Br. J. Sports Med.* **2011**, *45*, 553–558. [CrossRef] [PubMed]
11. Al Attar, W.S.A.; Soomro, N.; Sinclair, P.J.; Pappas, E.; Sanders, R.H. Effect of injury prevention programs that include the Nordic hamstring exercise on hamstring injury rates in soccer players: A systematic review and meta-analysis. *Sports Med.* **2017**, *47*, 907–916. [CrossRef]
12. Stergiou, M.; Calvo, A.L.; Forelli, F. Effectiveness of Neuromuscular Training in Preventing Lower Limb Soccer Injuries: A Systematic Review and Meta-Analysis. *J. Clin. Med.* **2025**, *14*, 1714. [CrossRef] [PubMed]
13. Steinberg, N.; Dar, G.; Dunlop, M.; Gaida, J.E. The relationship of hip muscle performance to leg, ankle and foot injuries: A systematic review. *Phys Sportsmed.* **2017**, *45*, 49–63. [CrossRef]
14. Soligard, T.; Myklebust, G.; Steffen, K.; Holme, I.; Silvers, H.; Bizzini, M.; Junge, A.; Dvorak, J.; Bahr, R.; Andersen, T.E. Comprehensive warm-up programme to prevent injuries in young female footballers: Cluster randomised controlled trial. *BMJ* **2008**, *337*, a2469. [CrossRef]
15. Thorborg, K.; Krommes, K.K.; Esteve, E.; Clausen, M.B.; Bartels, E.M.; Rathleff, M.S. Effect of specific exercise-based football injury prevention programmes on the overall injury rate in football: A systematic review and meta-analysis of the FIFA 11 and 11+ programmes. *Br. J. Sports Med.* **2017**, *51*, 562–571. [CrossRef]
16. Barboza, S.D.; Nauta, J.; Emery, C.; van Mechelen, W.; Gouttebarge, V.; Verhagen, E. A Warm-Up Program to Reduce Injuries in Youth Field Hockey Players: A Quasi-Experiment. *J. Athl. Train.* **2019**, *54*, 374–383. [CrossRef]
17. Weir, G.; Alderson, J.A.; Elliott, B.C.; Lee, S.; Devaprakash, D.; Starre, K.; Goodman, C.; Cooke, J.; Rechichi, C.; Armstrong, J.; et al. A 2-yr Biomechanically Informed ACL Injury Prevention Training Intervention in Female Field Hockey Players. *Transl. J. Am. Coll. Sports Med.* **2019**, *4*, 206–214. [CrossRef]
18. Myer, G.D.; Faigenbaum, A.D.; Ford, K.R.; Best, T.M.; Bergeron, M.F.; Hewett, T.E. When to Initiate Integrative Neuromuscular Training to Reduce Sports-Related Injuries and Enhance Health in Youth? *Curr. Sports Med. Rep.* **2011**, *10*, 155–166. [CrossRef] [PubMed]
19. Petushek, E.J.; Sugimoto, D.; Stoolmiller, M.; Smith, G.; Myer, G.D. Evidence-Based Best-Practice Guidelines for Preventing Anterior Cruciate Ligament Injuries in Young Female Athletes: A Systematic Review and Meta-analysis. *Am. J. Sports Med.* **2019**, *47*, 1744–1753. [CrossRef]
20. Parsons, J.L.; Alexander, M.J.L. Modifying Spike Jump Landing Biomechanics in Female Adolescent Volleyball Athletes Using Video and Verbal Feedback. *J. Strength Cond. Res.* **2012**, *26*, 1076–1084. [CrossRef] [PubMed]
21. Sugimoto, D.; Myer, G.D.; Barber Foss, K.D.; Pepin, M.J.; Micheli, L.J.; Hewett, T.E. Critical components of neuromuscular training to reduce ACL injury risk in female athletes: Meta-regression analysis. *Br. J. Sports Med.* **2016**, *50*, 1259–1266. [CrossRef] [PubMed]
22. Sugimoto, D.; Myer, G.D.; Foss, K.D.; Hewett, T.E. Dosage effects of neuromuscular training intervention to reduce anterior cruciate ligament injuries in female athletes: Meta- and sub-group analyses. *Sports Med.* **2014**, *44*, 551–562. [CrossRef]
23. Harry, J.R.; Lanier, R.; Nunley, B.; Blinch, J. Focus of attention effects on lower extremity biomechanics during vertical jump landings. *Hum. Mov. Sci.* **2019**, *68*, 102521. [CrossRef]
24. Donnelly, C.J.; Elliott, B.C.; Ackland, T.R.; Doyle, T.L.A.; Beiser, T.F.; Finch, C.F.; Cochrane, J.L.; Dempsey, A.R.; Lloyd, D.G. An anterior cruciate ligament injury prevention framework: Incorporating the recent evidence. *Res. Sports Med.* **2012**, *20*, 239–262. [CrossRef]

25. Hoffman, T.C.; Glasziou, P.P.; Boutron, I.; Milne, R.; Perera, R.; Moher, D.; Altman, D.G.; Barbour, V.; MacDonal, H.; Johnston, M.; et al. Better reporting of interventions: Template for intervention description and replication (TIDieR) checklist and guide. *BMJ* **2014**, *348*, g1687. [CrossRef] [PubMed]
26. Hewett, T.E.; Lindenfeld, T.N.; Riccobene, J.V.; Noyes, F.R. The Effect of Neuromuscular Training on the Incidence of Knee Injury in Female Athletes. *Am. J. Sports Med.* **1999**, *27*, 699–706. [CrossRef]
27. Hewett, T.E.; Myer, G.D.; Ford, K.R.; Heidt, R.S., Jr.; Colosimo, A.J.; McLean, S.G.; van den Bogert, A.J.; Paterno, M.V.; Succop, P. Biomechanical measures of neuromuscular control and valgus loading of the knee predict anterior cruciate ligament injury risk in female athletes: A prospective study. *Am. J. Sports Med.* **2005**, *33*, 492–501. [CrossRef] [PubMed]
28. Lephart, S.M.; Abt, J.P.; Ferris, C.M.; Sell, T.C.; Nagai, T.; Myers, J.B.; Irrgang, J.J. Neuromuscular and biomechanical characteristic changes in high school athletes: A plyometric versus basic resistance program. *Br. J. Sports Med.* **2005**, *39*, 932–938. [CrossRef]
29. Myer, G.D.; Ford, K.R.; Paterno, M.V.; Nick, T.G.; Hewett, T.E. The Effects of Generalized Joint Laxity on Risk of Anterior Cruciate Ligament Injury in Young Female Athletes. *Am. J. Sports Med.* **2008**, *36*, 1073–1080. [CrossRef]
30. Yoo, J.H.; Lim, B.O.; Ha, M.; Lee, S.W.; Oh, S.J.; Lee, Y.S.; Kim, J.G. A meta-analysis of the effect of neuromuscular training on the prevention of the anterior cruciate ligament injury in female athletes. *Knee Surg. Sports Traumatol. Arthrosc.* **2010**, *18*, 824–830. [CrossRef]
31. Zazulak, B.T.; Hewett, T.E.; Reeves, N.P.; Goldberg, B.; Cholewicki, J. Deficits in Neuromuscular Control of the Trunk Predict Knee Injury Risk: Prospective Biomechanical-Epidemiologic Study. *Am. J. Sports Med.* **2007**, *35*, 1123–1130. [CrossRef]
32. Benjaminse, A.; Gokeler, A.; Dowling, A.V.; Faigenbaum, A.; Ford, K.R.; Hewett, T.E.; Onate, J.A.; Otten, B.; Myer, G.D. Optimization of the anterior cruciate ligament injury prevention paradigm: Novel feedback techniques to enhance motor learning and reduce injury risk. *J. Orthop. Sports Phys.* **2015**, *45*, 170–182. [CrossRef]
33. Steib, S.; Rahlf, A.L.; Pfeifer, K.; Zech, A. Dose-Response Relationship of Neuromuscular Training for Injury Prevention in Youth Athletes: A Meta-Analysis. *Front. Physiol.* **2017**, *8*, 920. [CrossRef]
34. Borg, G.A. Psychophysical bases of perceived exertion. *Med. Sci. Sports Exerc.* **1982**, *14*, 377–381. [CrossRef] [PubMed]
35. Ford, K.R.; Myer, G.D.; Toms, H.E.; Hewitt, T.E. Gender Differences in the Kinematics of Unanticipated Cutting in Young Athletes. *Med. Sci. Sports Exerc.* **2005**, *37*, 124–129. [CrossRef]
36. Hermans, H.J.; Freriks, B.; Merletti, R.; Stegeman, D.; Blok, J.; Rau, G.; Disselhorst-Klug, C.; Hagg. *European Recommendations for Surface ElectroMyoGraphy*, 2nd ed.; Roessingh Research and Development: Enschede, The Netherlands, 1999.
37. Albertus-Kajee, Y.; Tucker, R.; Derman, W.; Lamberts, R.P.; Lambert, M.I. Alternative methods of normalizing EMG during running. *J. Electromyogr. Kinesiol.* **2011**, *21*, 579–586. [CrossRef] [PubMed]
38. Burden, A.M. How should we normalize electromyograms obtained from healthy participants What we have learned from over 25 years of research. *J. Electromyogr. Kinesiol.* **2010**, *20*, 1023–1035. [CrossRef]
39. Winter, D.A. *Biomechanics and Motor Control of Human Movement*, 3rd ed.; Wiley: Hoboken, MJ, USA; Chichester, UK, 2005; ISBN 047144989X.
40. Roewer, B.D.; Ford, K.R.; Myer, G.D.; Hewett, T.E. The 'impact' of force filtering cut-off frequency on the peak knee abduction moment during landing: Artefact or 'artifiction'? *Br. J. Sports Med.* **2014**, *48*, 464–468. [CrossRef] [PubMed]
41. Pappas, E.; Shiyko, M.P.; Ford, K.R.; Myer, G.D.; Hewett, T.E. Biomechanical Deficit Profiles Associated with ACL Injury Risk in Female Athletes. *Med. Sci. Sports Exerc.* **2016**, *48*, 107–113. [CrossRef]
42. Neto, W.K.; Soares, E.G.; Vieira, T.L.; Aguiar, R.; Chola, T.A.; Sampaio, V.L.; Gama, E.F. Gluteus Maximus Activation During Common Strength and Hypertrophy Exercises: A Systematic Review. *JSSM* **2020**, *19*, 195–203.
43. Boren, K.; Conrey, C.; Le Coguic, J.; Paprocki, L.; Voight, M.; Robinson, T.K. Electromyographic analysis of gluteus medius and gluteus maximus during rehabilitation exercises. *IJSPT* **2011**, *6*, 206–223.
44. Khayambashi, K.; Ghoddosi, N.; Straub, R.K.; Powers, C.M. Hip muscle strength predicts non-contact anterior cruciate ligament injury in male and female athletes: A prospective study. *Am. J. Sports Med.* **2016**, *44*, 355–361. [CrossRef] [PubMed]
45. Wilderman, D.R.; Ross, S.E.; Padua, D.A. Thigh Muscle Activity, Knee Motion, and Impact Force During Side-Step Pivoting in Agility-Trained Female Basketball players. *J. Athl. Train.* **2009**, *44*, 14–25. [CrossRef]
46. van den Tillaar, R.; Solheim, J.A.B.; Bencke, J. Comparison of Hamstring Muscle Activation during High-Speed Running and various Hamstring Strengthening Exercises. *Int. J. Sports Phys. Ther.* **2017**, *12*, 718–727. [CrossRef] [PubMed]
47. van Dyk, N.; Behan, F.P.; Whiteley, R. Including the Nordic hamstring exercise in injury prevention programmes halves the rate of hamstring injuries: A systematic review and meta-analysis of 8459 athletes. *Br. J. Sports Med.* **2019**, *53*, 1362–1370. [CrossRef]
48. Zebis, M.K.; Benke, J.; Andersen, L.L.; Dossing, S.; Alkaer, T.; Magnusson, P.; Kjaer, M.; Aagaard, P. The Effects of Neuromuscular Training on Knee Joint Motor Control During Sidecutting in Female Elite Soccer and Handball Players. *J. Clin. Med.* **2008**, *18*, 329–337. [CrossRef] [PubMed]
49. Zebis, M.K.; Andersen, L.L.; Brandt, M.; Myklebust, G.; Bencke, J.; Lauridsen, H.B.; Bandholm, T.; Thorborg, K.; Hölmich, P.; Aagaard, P. Effects of evidence-based prevention training on neuromuscular and biomechanical risk factors for ACL injury in adolescent female athletes: A randomized controlled trial. *Br. J. Sports Med.* **2016**, *50*, 552–557. [CrossRef]

50. Gentil, P.; Souza, D.; Santana, M.; Alves, R.R.; Campos, M.H.; Pinto, R.; Bottaro, M. Multi- and Single-Joint Resistance Exercises Promote Similar Plantar Flexor Activation in Resistance Trained Men. *Int. J. Environ. Res. Public Health* **2020**, *17*, 9487. [CrossRef]
51. Ali, N.; Andersen, M.S.; Rasmussen, J.; Robertson, D.G.E.; Rouhi, G. The application of musculoskeletal modeling to investigate gender bias in non-contact ACL injury rate during single-leg landings. *CMBBE* **2014**, *17*, 1602–1616. [CrossRef]
52. Benke, J.; Curtis, D.; Krogshede, C.; Jensen, L.K.; Bandholm, T.; Zebis, M.K. Biomechanical evaluation of the side-cutting manoeuvre associated with ACL injury in young female handball players. *Knee Surg. Sports Traumatol. Arthrosc.* **2013**, *21*, 1876–1881. [CrossRef]
53. Błaszczyszyn, M.; Konieczny, M.; Pakosz, P. Analysis of ankle sEMG on both stable and unstable surfaces for elderly and young women—A pilot study. *Int. J. Environ. Res. Public Health* **2019**, *16*, 1544. [CrossRef]
54. Brunner, R.; Friesenbichler, B.; Casartelli, N.C.; Bizzini, M.; Maffiuletti, N.A.; Niedermann, K. Effectiveness of multicomponent lower extremity injury prevention programmes in team-sport athletes: An umbrella review. *Br. J. Sports Med.* **2019**, *53*, 282–288. [CrossRef]
55. Timmins, R.G.; Opar, D.A.; Williams, M.D.; Schache, A.G.; Dear, N.M.; Shield, A.J. Reduced biceps femoris myoelectrical activity influences eccentric knee flexor weakness after repeat sprint running. *Scand. J. Med. Sci. Sports* **2014**, *24*, e299–e305. [CrossRef]
56. Bakker, R.; Tomescu, S.; Brenneman, E.; Hangalur, G.; Laing, A.; Chandrashekar, N. Effect of sagittal plane mechanics on ACL strain during jump landing. *J. Orthop.* **2016**, *34*, 1636–1644. [CrossRef]
57. Ireland, M.L.; Willson, J.D.; Ballantyne, B.T.; McClay-Davis, I. Hip Strength in Females With and Without Patellofemoral Pain. *J. Orthop. Sports Phys.* **2003**, *33*, 671–676. [CrossRef]
58. Baldon Rde, M.; Serrão, F.V.; Scattone Silva, R.; Piva, S.R. Effects of functional stabilization training on pain, function, and lower extremity biomechanics in women with patellofemoral pain: A randomized clinical trial. *J. Orthop. Sports Phys. Ther.* **2014**, *44*, 240–251+A1–A8. [CrossRef] [PubMed]
59. Quatman, C.E.; Ford, K.R.; Myer, G.D. Maturation Leads to Gender Differences in Landing Force and Vertical Jump Performance, A Longitudinal Study. *Am. J. Sports Med.* **2006**, *34*, 806–813. [CrossRef]
60. Myer, G.D.; Ford, K.R.; Khoury, J.; Succop, P.; Hewett, T.E. Development and Validation of a clinical-Based Prediction Tool to Identify Female Athletes at High Risk for Anterior Cruciate Ligament Injury. *Am. J. Sports Med.* **2010**, *38*, 2025–2033. [CrossRef] [PubMed]
61. Ireland, M.L. The female ACL: Why is it more prone to injury. *J. Orthop.* **2016**, *13*, A1–A4. [CrossRef]
62. Della Villa, F.; Buckthorpe, M.; Grassi, A.; Nabiuzzi, A.; Tosarelli, F.; Zaffagnini, S.; Della Villa, S. Systematic video analysis of ACL injuries in professional male football (soccer): Injury mechanisms, situational patterns and biomechanics study on 134 consecutive cases. *Br. J. Sports Med.* **2020**, *54*, 1423–1432. [CrossRef] [PubMed]
63. Hopper, A.J.; Haff, E.E.; Joyce, C.; Lloyd, R.S.; Haff, G.G. Neuromuscular Training Improves Lower Extremity Biomechanics Associated with Knee Injury during Landing in 11–13 Year Old Female Netball Athletes: A Randomized Control Study. *Front. Physiol.* **2017**, *8*, 883. [CrossRef]
64. Yom, J.P.; Owens, T.; Arnett, S.; Beebe, J.; Son, V. The effects of an unanticipated side-cut on lower extremity kinematics and ground reaction forces during a drop landing. *Sports Biomech.* **2018**, *18*, 414–425. [CrossRef]
65. Woo, S.L.; Hollis, J.M.; Adams, D.J.; Lyon, R.M.; Takai, S. Tensile properties of the human femur-anterior cruciate ligament-tibia complex. The effects of specimen age and orientation. *Am. J. Sports Med.* **1991**, *19*, 217–225. [CrossRef] [PubMed]
66. Distefano, L.J.; Blackburn, J.T.; Marshall, S.W.; Padua, D.A. Gluteal Activation During Common Therapeutic Exercises. *J. Orthop. Sports Phys.* **2009**, *39*, 532–540. [CrossRef] [PubMed]
67. Irmischer, B.S.; Harris, C.; Pfeiffer, R.P.; DeBeliso, M.A.; Adams, K.J.; Shea, K.G. Effects of a knee Ligament Injury Prevention Exercise Program on Impact Forces in Women. *J. Strength Cond. Res.* **2004**, *18*, 703–707. [CrossRef] [PubMed]
68. Marinsek, M. Basic Landing Characteristics and their Application in Artistic Gymnastics. *Sci. Gymnast. J.* **2010**, *2*, 59–67. [CrossRef]

Disclaimer/Publisher's Note: The statements, opinions and data contained in all publications are solely those of the individual author(s) and contributor(s) and not of MDPI and/or the editor(s). MDPI and/or the editor(s) disclaim responsibility for any injury to people or property resulting from any ideas, methods, instructions or products referred to in the content.

bioengineering

MDPI

Article

Kinematic Alterations with Changes in Putting Distance and Slope Incline in Recreational Golfers

Shawn M. Robbins [1,2,*], Philippe Renaud [3] and Ukadike Chris Ugbolue [4,5]

1 School of Physical and Occupational Therapy, McGill University, Montreal, QC H3G 1Y5, Canada
2 Centre for Interdisciplinary Research in Rehabilitation and Lethbridge-Layton-MacKay Rehabilitation Centre, Montreal, QC H4B 1T3, Canada
3 Department of Kinesiology and Physical Education, McGill Research Centre for Physical Activity and Health, McGill University, Montreal, QC H2W 1S4, Canada; philjrenaud@hotmail.com
4 Biomechanics Laboratory, Division of Sport and Exercise, School of Health and Life Sciences, University of the West of Scotland, Paisley PA1 2BE, UK; u.ugbolue@uws.ac.uk
5 Faculty of Sports Science, Ningbo University, Ningbo 315211, China
* Correspondence: shawn.robbins@mcgill.ca; Tel.: +1-514-396-2713

Abstract: Golfers must modify their motor patterns when the demands of a putting task change. The objective was to compare joint angles and putter kinematics during putting at two distances and inclines. Recreational golfers (n = 14) completed putts over four conditions: 3-foot putts on flat and incline surfaces, and 7-foot putts on flat and incline surfaces. A Vicon motion capture system measured kinematic data. Joint angles, putter angles, and spatiotemporal variables were calculated. Analysis of variance compared spatiotemporal variables, and statistical parametric mapping compared angles between putts. There were faster putter head and ball velocities during longer and incline putts. The amplitude and time of backswing increased with longer putts. Longer putts resulted in increased trunk axial rotation during backswing, downswing, and follow-through, while incline putts only resulted in greater rotation during follow-through. There were minimal differences in shoulder angle. There was greater head rotation toward the hole during all putting phases for longer putts and during follow-through for incline putts. The trunk is the primary mechanism to increase putter head amplitude, and thereby velocity, when putting from longer distances. A similar strategy could be used when putting uphill. Additional work should confirm these results in highly skilled golfers.

Keywords: golf; putting; motion capture; kinematics; biomechanics

1. Introduction

The game of golf is popular throughout the world, with an estimated 45 million golf participants in the United States [1] and 62.3 million in the rest of the world [2]. Golf presents a unique opportunity to study motor learning principles as the outcome of a golf stroke is simple (i.e., accuracy of the shot); it is played in a relatively controlled environment, and there are variations in the task demands [3]. One aspect to study is variation in motor patterns in response to changing task demands, including golfers modifying their body kinematics when putting from different distances or using different clubs. The success of motor learning training strategies (e.g., augmented feedback, cognitive training, explicit learning, and internal focus of attention) at improving outcomes and motor patterns can also be examined in golfers [3]. Considering the complexity of the golf swing, numerous studies have used biomechanical tools to evaluate and quantify the motor patterns or

https://doi.org/10.3390/bioengineering12010069

kinematics of this activity [4]. Most studies have examined the full golf swing, with golfers using drivers or wedges. Less frequently studied is the putting stroke.

On the Professional Golfers Association Tour in 2024, the average putts per round and total strokes per round were 29 and 71, respectively [5]. Thus, putting accounted for 40% of all strokes. For the current study, the putter refers to the club used in the putting stroke, and the putter head refers to the part of the club that contacts the ball. The putting stroke has been modelled as a double pendulum, which includes the putter and both arms [6,7]. To accomplish this motion, the "shoulder is meant to roll in an up-and-down fashion, and the two hands hold the putter together, moving back and forth to putt symmetrically" [8]. The putter face should remain perpendicular to the initial ball direction line [9,10]. The ball should also be hit on the upward stroke, which increases topspin, improving the efficiency of the ball roll and reducing skidding [10]. Head movement during putting is considered poor technique, and the golfer's head should remain stationary [11].

Golfers putt from various distances, and adapting to different distances can be challenging. Typically, golfers should increase the amplitude of their stroke when putting from longer distances, while maintaining a consistent tempo [12,13]. In support, previous research has demonstrated increased amplitude of the putter head during the backswing, downswing, and follow-through phases of the putting stroke when putting from longer distances in highly skilled golfers [14–16]. The duration of backswing and downswing also increased with longer distances [14,15]. This increased backswing amplitude will naturally increase the putter head velocity, and faster putter head velocity has been demonstrated with longer putts [14–17]. Putter head velocity has been shown to be greater in people with no putting experience compared to expert golfers [14], although these difference were not present when comparing recreational to expert golfers [11].

Most studies have examined putter kinematics, and there has been limited research examining body kinematics during putting [11]. One study found that expert golfers had lower displacements of their heads compared to less skilled golfers during putting, although the results were not statistically significant [11]. This indicates that the head should remain stable when putting. In expert golfers, there were differences in thorax and lumbar spine angles between proficient (>79% putting success) and non-proficient (<79% putting success) golfers when completing 2 m putts [6]. Research is required to examine how kinematics of the golfer's body change between putting distances. We know that putter kinematics change with increasing distances, but it is not clear how golfers produce these changes by modifying their joint angles. Furthermore, other task constraints have not been examined, including comparing putting between flat and inclined surfaces. Therefore, the objective was to compare joint angles, putter angles, and timing variables during putting at different distances and surface inclines in recreational golfers. It was hypothesized that there would be increased amplitudes in putter and shoulder angles, and faster ball and putter head velocities with longer distance and incline slopes.

2. Materials and Methods

2.1. Participants

This cross-sectional study was a secondary analysis of an existing database. The sample included 14 male recreational golfers. All golfers were right-handed, and all had experience playing golf. Exclusion criteria included current back, wrist, elbow, or arm injuries and additional injuries/diseases that would prevent them from playing golf. The study was approved by the School of Health and Life Sciences Ethics Committee at the University of the West of Scotland (#5-3-14-002). All participants provided written, informed consent.

Data were collected at the Biomechanics Laboratory at the University of the West of Scotland (Lanarkshire campus). Participants attended one data collection session. Demographic information was collected (e.g., age). The handicap and golfing experience (e.g., frequency of play, how long they had been playing) of the participants were not known.

2.2. Putting Task

Participants completed putts on an artificial mat (SKLZ Accelerator Pro Putting Mat, Carlsbad, CA, USA), which was attached to a wooden board (8 feet long, 1 foot wide). Participants used a standard putter (Classic Smoke Daytona 79 Putter, 34-inch shaft length, TaylorMade, Basingstoke, UK) and ball (Pro V1 golf ball, Titleist, Cambridgeshire, UK). During putting (Figure 1), participants stood on wooden blocks (30 × 10 cm) so they were level with the putting surface. Participants completed putts over four conditions: 3-foot flat, 7-foot flat, 3-foot incline, and 7-foot incline. To create the incline condition, extra wooden blocks were placed at the end of the putting surface and under the left foot of the participants to create a 5.7° incline. Participants were permitted a 5 min warm-up to become familiar with the equipment. They then completed three trials for each condition, and the order was randomized. They were instructed to putt the ball in the hole, and no additional instructions about technique were provided. Both successful and unsuccessful putts were included.

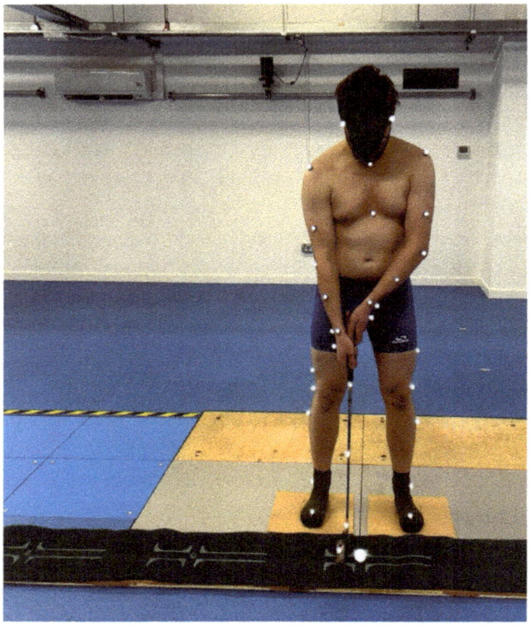

Figure 1. Data collection setup. During testing, we ensured the artificial putting surface was flat with no wrinkles.

2.3. Biomechanical Data Collection

Kinematic data were recorded with an eight-camera motion capture system (Bonita cameras, Vicon Metrics Ltd., Oxford, UK) at 250 Hz. Vicon Nexus software (version 2.11, Vicon Metric Ltd., Oxford, UK) was used to record the data. Retroreflective markers (14 mm) were placed on the participants according to the Plug-in Gait marker system [18]. Four markers were placed on the putter, including three on the shaft (top, mid, and bottom) and one on the heel of the putter head. The golf ball was covered in retroreflective tape to ensure that it was visible during the data capture trial sessions. Additional 14 mm markers were placed on the spinal processes (C7 to L5) but were not used in the current study. In total, 55 markers were placed on the participant and putter. Relevant anthropometric measures were taken first, including height, weight, leg length, joint widths (knee, elbow, wrist, and hand), and shoulder offset. Participants then completed a static standing trial before completing the putting conditions.

2.4. Data Processing

Data were processed with Visual3d (version 2024.05.3, HAS Motion, Kingston, ON, Canada). Maker data were filtered with a lowpass, 4-order Butterworth filter with a 5 Hz cutoff [17]. The segments were defined according to the Plug-In Gait model [19]. Joint angles were calculated for the head (with respect to the trunk), trunk (with respect to the pelvis), and shoulder from the trail (right) side (arm with respect to the trunk) using Euler XYZ, YXZ, and ZYZ sequences, respectively, consistent with either previous golf studies (trunk) or ISB recommendations (shoulder) [20–22]. Data for the left shoulder were not consistently available and were not presented. Joint angles included in statistical analyses were head and trunk extension/flexion (positive = extension), sideflexion (positive = right sideflexion), and axial rotation (positive = left rotation). For the shoulder, plane of elevation (0° = horizontal abduction, 90° = forward flexion) and elevation (negative = elevation) were analyzed. These angles were considered since they would likely contribute to the putting stroke or could represent an error in putting (e.g., head rotation). Additionally, the putter angle relative to the lab coordinate system was determined using an Euler XYZ sequence. The rotation of the putter angle about the Y-axis was considered, with positive values representing putter head rotation toward the hole and negative values rotation away from the hole.

The putting stroke was divided into three phases: backswing, downswing, and follow-through [17,23]. Backswing began when the putter head marker velocity started moving in the negative direction (away from the hole) until the putter head marker reached its position furthest from the hole. Downswing then began and concluded when the ball reflective tape velocity started to increase, representing ball impact. Follow-through began at ball impact and concluded when the putter head marker velocity switched from a positive value (toward the hole) to a negative value. The time of each putting phase was determined, along with the peak ball and putter head velocity. Joint and putter angle waveforms were time-normalized to 100% of the putting stroke (backswing to follow-through). The three trials from each condition were then averaged for each participant.

2.5. Statistical Analysis

Since this was a secondary analysis of an existing database, a formal sample size calculation was not conducted. Descriptive statistics were calculated for the study variables. Two-way repeated analysis of variance (ANOVA) compared peak ball velocity, putter head velocity, and the times of each phase between the distance (3 vs. 7 feet) and incline (flat vs. incline) conditions. ANOVAs were conducted using SPSS (version 24, IBM Corp., Armonk, NY, USA). Statistical parametric mapping (SPM) two-way repeated measures ANOVAs compared joint and putter angle waveforms between distance and incline conditions [24]. The time node was each 1% of the putting stroke. A critical threshold was determined, and F-statistics that surpassed this critical threshold were deemed statistically significant ($p < 0.05$). The SPM analysis was completed with the spm1D toolbox in Matlab (version 2018a, MathWorks Inc., Natick, MA, USA) [25,26].

3. Results

3.1. Ball and Putter Head Velocity

There were statistically significant differences in peak ball velocity between distance ($p < 0.001$) and incline ($p < 0.001$) conditions, although the interaction was not significant ($p = 0.337$). Likewise, there were statistically significant differences in peak putter head velocity between distance ($p < 0.001$) and incline ($p < 0.001$) conditions, and the interaction was not significant ($p = 0.200$). There were faster peak ball and putter head velocities during the 7-foot and incline putts (Table 1).

Table 1. Demographic statistics and peak ball and putter head velocity for the study sample.

Variable		Mean (Standard Deviation)
Age (y)		22 (12)
Mass (kg)		77.86 (13.12)
Height (m)		1.78 (0.06)
Body mass index (kg/m^2)		24.63 (3.42)
Peak ball velocity (m/s)	3-foot flat	1.29 (0.08)
	3-foot incline	1.68 (0.22)
	7-foot flat	2.08 (0.35)
	7-foot incline	2.57 (0.29)
Peak putter head velocity (m/s)	3-foot flat	0.93 (0.07)
	3-foot incline	1.16 (0.12)
	7-foot flat	1.33 (0.14)
	7-foot incline	1.62 (0.16)

3.2. Phase Times

There were significant differences in backswing times between distances ($p = 0.001$), but incline ($p = 0.834$) and interaction ($p = 0.416$) effects were not significant (Table 2). Backswing times were longer during 7-foot compared to 3-foot putts (mean difference = 0.06 s). There were no significant differences in downswing times for distance ($p = 0.614$), incline ($p = 0.065$), and interaction ($p = 0.710$) effects (Table 2). There were significant differences in follow-through times for the distance effect ($p = 0.042$), but incline ($p = 0.736$) and interaction ($p = 0.737$) effects were not significant (Table 2). Follow-through times were longer during 7-foot compared to 3-foot putts (mean difference = 0.03 s).

Table 2. Mean (standard deviation) for each putting phase for the four conditions.

Condition	Backswing	Downswing	Follow-Through
3-foot flat (s)	0.53 (0.07)	0.25 (0.06)	0.52 (0.20)
3-foot incline (s)	0.53 (0.08)	0.24 (0.05)	0.51 (0.17)
7-foot flat (s)	0.58 (0.09)	0.25 (0.05)	0.54 (0.15)
7-foot incline (s)	0.59 (0.10)	0.24 (0.05)	0.55 (0.20)

3.3. Putter and Joint Angles

Significant SPM distance and incline effects are discussed below. There were no significant interactions. There were greater putter angles away from the hole (negative values) during the second half of backswing and early downswing (28 to 58% putting stroke) and greater angles toward the hole (positive values) during follow-through (74 to 100%) for 7-foot compared to 3-foot putts (Figure 2). Additionally, there were greater angles toward the hole during follow-through (75 to 100%) for incline compared to flat putts. There was also greater trail-side shoulder elevation (more negative values) at the end of follow-through for 7-foot compared to 3-foot putts (Figure 2).

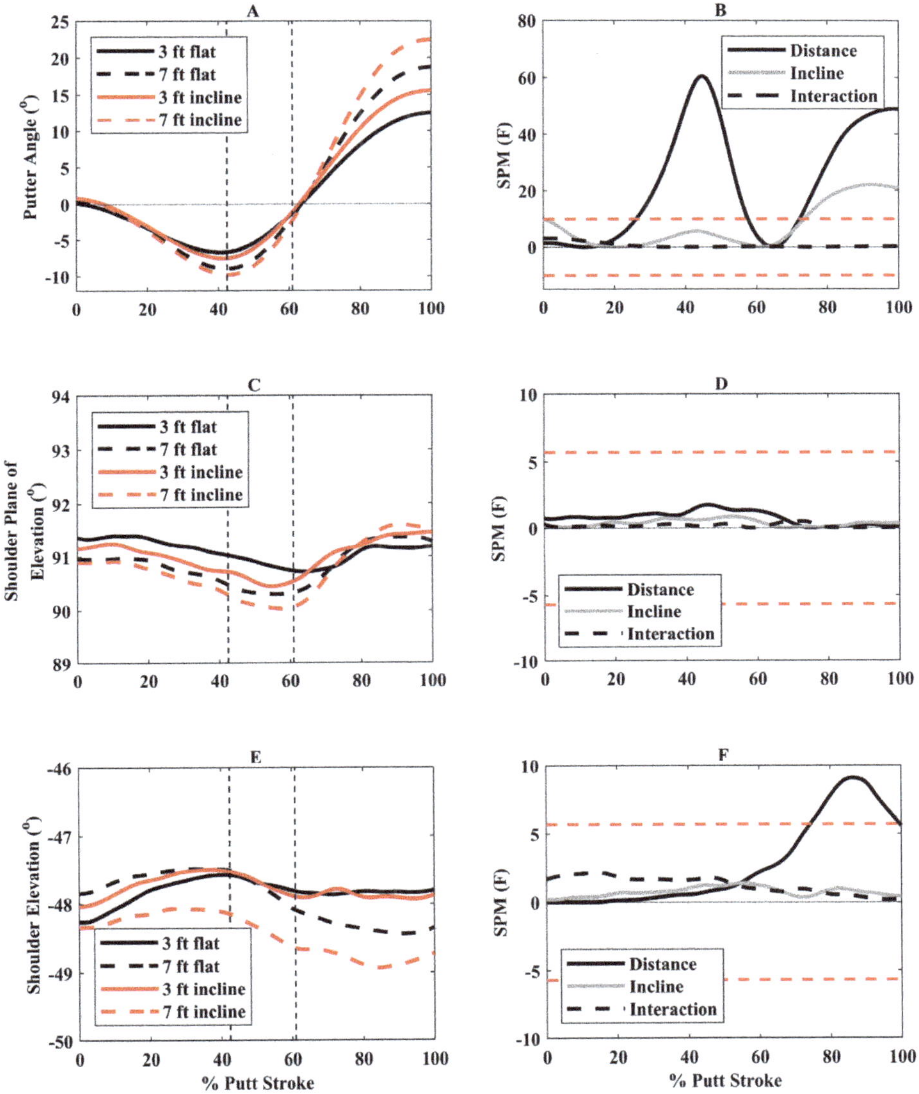

Figure 2. (**A**) Putter angles (positive = putter rotation toward hole), (**C**) Trail shoulder plane of elevation angles (0° = horizontal abduction, 90° = forward flexion), and (**E**) Trail shoulder elevation angles (negative = elevation) for the four putting conditions over the putting stroke (0° = start of backswing, 100° = end of follow-through). The vertical dashed lines represent the average transition between backswing and downswing (42.4% of putting stroke) and between downswing and follow-through (60.9% of putting stroke). Associated statistical parametric mapping (SPM) (F) plots are in panels (**B**,**D**,**F**). If the scalar statistic for each effect (distance, incline, interaction) goes above or below the critical threshold (red dashed lines), this indicates significant differences between putting conditions for that specific effect.

The 7-foot putts resulted in greater extension of the participants' heads throughout the putt (0 to 100%) and greater rotation toward the hole (left rotation) during parts of backswing, downswing, and follow-through (26 to 69% and 79 to 100% putting stroke) (Figure 3) compared to 3-foot putts. Also, there was greater rotation of the participants' heads toward the hole at the end of the backswing (94 to 100%) for incline compared to flat putts (Figure 3).

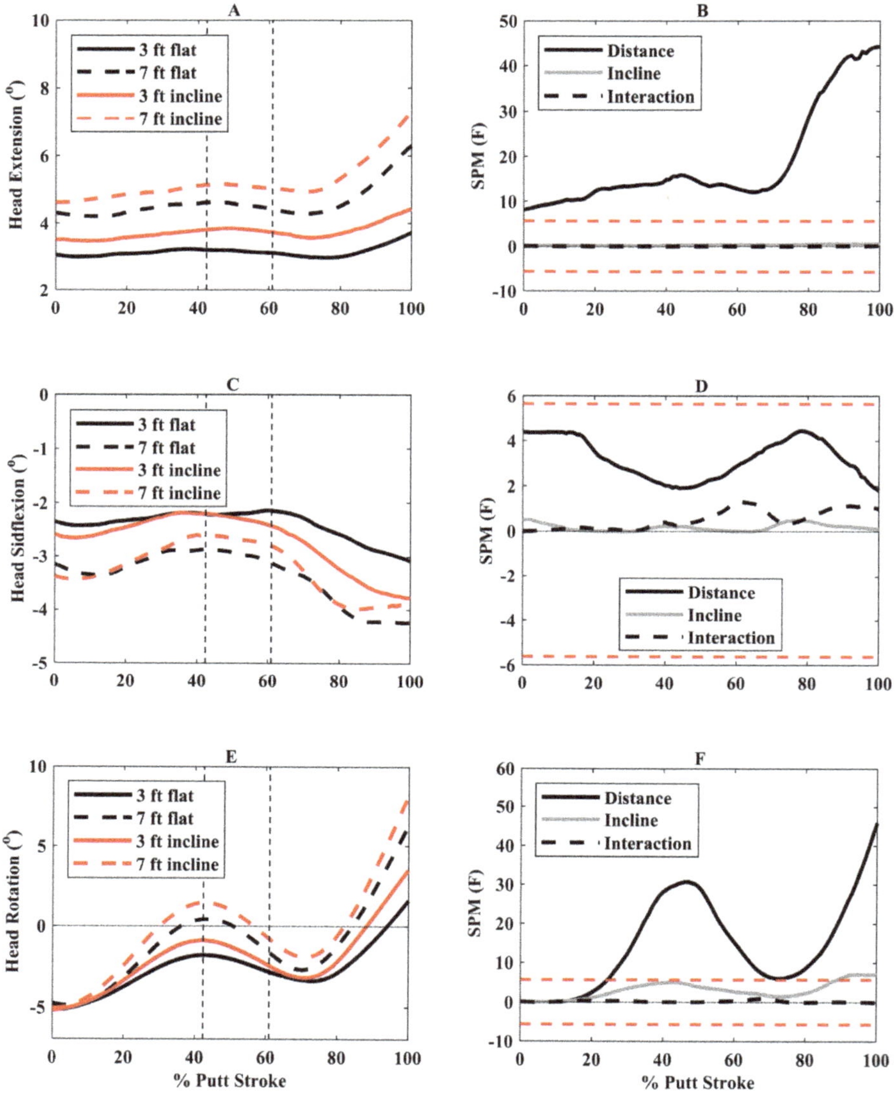

Figure 3. (**A**) Head extension angles (positive = extension), (**C**) Head sideflexion angles (positive = right sideflexion), and (**E**) Head axial rotation angles (positive = left rotation) for the four putting conditions over the putting stroke (0° = start of backswing, 100° = end of follow-through). The vertical dashed lines represent the average transition between backswing and downswing (42.4% of putting stroke) and between downswing and follow-through (60.9% of putting stroke). Associated statistical parametric mapping (SPM) (F) plots are in panels (**B**,**D**,**F**). If the scalar statistic for each effect (distance, incline, interaction) goes above or below the critical threshold (red dashed lines), this indicates significant differences between putting conditions for that specific effect.

The 7-foot putts resulted in greater trunk sideflexion away from the hole (right sideflexion) during follow-through (69 to 100%) compared to 3-foot putts. There was greater axial trunk rotation away from the hole (right rotation) during backswing/downswing (25 to 59%) and toward the hole (left rotation) during follow-through (75 to 100%) for 7-foot compared to 3-foot putts (Figure 4).

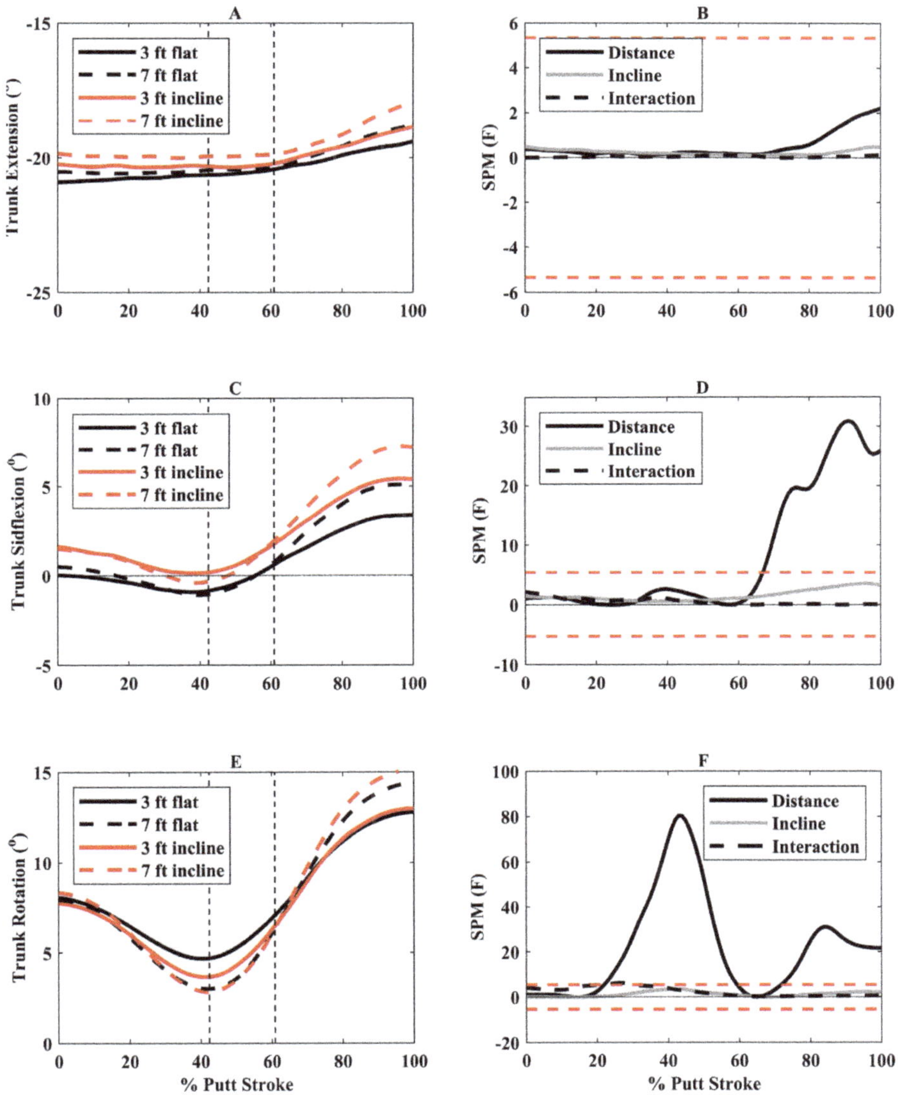

Figure 4. (**A**) Trunk extension angles (positive = extension), (**C**) Trunk sideflexion angles (positive = right sideflexion), and (**E**) Trunk axial rotation angles (positive = left rotation) for the four putting conditions over the putting stroke (0° = start of backswing, 100° = end of follow-through). The vertical dashed lines represent the average transition between backswing and downswing (42.4% of putting stroke) and between downswing and follow-through (60.9% of putting stroke). Associated statistical parametric mapping (SPM) (F) plots are in panels (**B**,**D**,**F**). If the scalar statistic for each effect (distance, incline, interaction) goes above or below the critical threshold (red dashed lines), this indicates significant differences between putting conditions for that specific effect.

4. Discussion

This study was novel since there is limited data evaluating changes in body kinematics when putting from different distances and incline conditions. The participants modified their movement patterns in response to the varying task demands. When putting from longer distances, recreational golfers increased the amplitude and time of their backswing, thereby increasing putter head velocity. This was likely accomplished by an increase in trunk axial rotation with longer putts, with minimal changes in shoulder angle. When putting on an incline, there was an increase in putter head velocity without an increase in putter head amplitude during backswing. These findings indicate that the trunk is a mechanism to increase putter head amplitude, and thereby velocity, when putting from longer distances. A similar approach should be considered when putting uphill. Results should be interpreted cautiously since participants were recreational golfers, and results cannot be generalized to professional golfers.

Consistent with our hypothesis, participants increased putter angle amplitudes during backswing, leading to longer backswing times, when putting from longer distances. This resulted in faster putter head and ball velocities. The follow-through putter angle amplitude and time also increased due to the faster putter head velocities. However, there was no change in the downswing times. In support, previous research demonstrated an increase in putter head amplitude and velocity, and in backswing times when putting from longer distances, in golfers with varying skill levels from novice to highly skilled [13–17,27]. Thus, coaches should encourage golfers to increase their putter angle amplitudes during backswing when golfers are adjusting to longer putts. For incline putts, there were faster putter head and ball velocities compared to the flat putts. Inconsistent with our hypothesis, there was no increase in the putter angle amplitude during backswing or the time of this phase. This finding likely indicates that participants used other mechanisms to increase their putter head velocity, such as swinging the putter harder, when putting on the incline surface. From a performance perspective, recreational golfers should also increase their putter angle amplitudes during backswing to have the necessary putter head velocity when putting uphill. The current findings might not be generalizable to highly skilled golfers as differences in putting kinematics have been found between skill levels [10,14,17,28]. Future research should evaluate how highly skilled golfers adapt to sloped surfaces, and if training intervention using motor control principles can help golfers of all skill levels adapt to varying putting distances and slopes.

To perform longer putts, participants increased their trunk axial rotation. This included greater axial rotation away from the hole (right rotation) during backswing and toward the hole during follow-through (left rotation) during longer putts (Figure 4). This motion would increase putter angle, thereby increasing putter head velocity, allowing them to putt further. Research investigating golf swings with drivers and irons also found that increased trunk axial rotation was related to clubhead speed [29]. For incline putts, there were small, nonsignificant increases in trunk axial rotation (Figure 4), and thus other mechanisms likely contributed to the faster putter head velocities during incline putts. We expected increased trail shoulder angle amplitudes with longer and incline putts. However, there was only a small increase in trail shoulder elevation during follow-through for longer putts, with differences generally being around 1°. Thus, increased putter angle amplitudes and faster putter head velocities during longer putts are more likely generated by trunk axial rotation. Longer distances also resulted in greater trunk right sideflexion (away from the hole) during follow-through, which may help maintain the centre of mass within the base of support as the putter swings in the opposite direction (toward the hole). Few studies have examined trunk and shoulder angles during putting, and thus it is difficult to locate our findings within the literature. One study found greater thorax and lumbar motions in the frontal plane (sideflexion) in proficient putters, while nonproficient putters had greater thorax and lumbar flexion [6]. Our study provides preliminary evidence that golfers should increase their axial trunk rotation to increase putter head velocity when putting from longer distances. However, more research is needed to examine how the body, and not just the putter, adapts to different putting conditions, especially in highly skilled golfers.

The position of the participants' heads was also impacted by putting conditions. The heads of the participants were more extended throughout the putting stroke for longer distances. However, this was less than 2° generally, except differences were larger at the end of follow-through (Figure 3). There was greater left rotation (toward the hole) of the participants' heads during parts of backswing and downswing, and at the end of follow-through for longer putts. This latter finding was also present when putting from an incline compared to a flat surface. These findings indicate that participants lifted their heads when putting from longer distances, which is considered poor putting technique [11]. Lifting the head near the end of follow-through is likely not a concern. Hower, head rotation during backswing and downswing could negatively impact longer distance putts by altering body position and, ultimately, putter path and should be discouraged in golfers. The current sample consisted of recreational golfers, and these findings may vary in professional golfers. In support, a previous study of expert golfers (n = 5) and less skilled golfers (n = 11) found that expert golfers had less head displacement during putting, although the results were not statistically significant [11].

This study had several limitations. The success of the putts was not recorded and thus relationships between body and putter kinematics with putting success could not be explored. All participants were recreational male golfers, and their handicaps and golfing experience (e.g., frequency) were not known. These golfers were included since their body kinematics were available in the database. Results cannot be generalized to highly skilled/professional golfers or female golfers. The sample size of the database was small and should be increased in future studies. There were not sufficient marker data to calculate left shoulder, bilateral elbow, or bilateral wrist angles. Their role in the putting stroke should be considered. Although putter and body kinematics are important when evaluating putting performance, other factors should also be considered in future studies, including the ability of golfers to read the green [9].

5. Conclusions

In conclusion, recreational golfers changed their motor patterns in response to different task demands. They increased the amplitude of their backswing, resulting in faster putter head velocities when putting from longer distances. Increases in trunk axial rotation were likely the primary mechanism to accomplish this greater backswing amplitude. When putting uphill, participants increased their putter head velocity without a significant increase in backswing amplitude or trunk axial rotation. Thus, they relied on other mechanisms to meet the required velocity demands of putting uphill. Future research should examine more skilled golfing populations (e.g., professionals) and female golfers, examine kinematics in more joints, and explore relationships between body kinematics and putting success.

Author Contributions: Conceptualization, S.M.R. and U.C.U.; methodology, U.C.U.; software, S.M.R. and P.R.; validation, U.C.U.; formal analysis, S.M.R. and P.R.; investigation, U.C.U.; resources, U.C.U.; data curation, S.M.R. and U.C.U.; writing—original draft preparation, S.M.R.; writing—review and editing, S.M.R., P.R. and U.C.U.; visualization, S.M.R., P.R. and U.C.U.; supervision, U.C.U.; project administration, U.C.U.; funding acquisition, U.C.U. All authors have read and agreed to the published version of the manuscript.

Funding: This research received no external funding.

Institutional Review Board Statement: The study was conducted in accordance with the Declaration of Helsinki and approved by the School of Health and Life Sciences Ethics Committee at the University of the West of Scotland (#5-3-14-002, January 2017).

Informed Consent Statement: Informed consent was obtained from all subjects involved in the study.

Data Availability Statement: Data are not available due to privacy concerns.

Acknowledgments: The authors would like to thank Luke Cerexhe, Kyle Finnie, Robert Brown, and Scott Alexander for their technical support.

Conflicts of Interest: The authors declare no conflicts of interest.

References

1. National Golf Foundation. Golf Participation Update—Bigger, Younger and Cooler. Available online: https://www.ngf.org/golf-participation-update-bigger-younger-and-cooler/ (accessed on 18 November 2024).
2. The Royal and Ancient Golf Club of St. Andrews. *Global Golf Participation 2024*; The Royal & Ancient Golf Club of St Andrews: St. Andrews/Scotland, UK, 2024.
3. Barzyk, P.; Gruber, M. Motor learning in golf—A systematic review. *Front. Sports Act. Living* **2024**, *6*, 1324615. [CrossRef] [PubMed]
4. Bourgain, M.; Rouch, P.; Rouillon, O.; Thoreux, P.; Sauret, C. Golf swing biomechanics: A systematic review and methodological recommendations for kinematics. *Sports* **2022**, *10*, 91. [CrossRef] [PubMed]
5. Professional Golfer's Association Tour. Available online: https://www.pgatour.com/stats (accessed on 18 November 2024).
6. Delphinus, E.M.; Sayers, M.G.L. Putting proficiency: Contributions of the pelvis and trunk. *Sports Biomech.* **2012**, *11*, 212–222. [CrossRef] [PubMed]
7. Munzert, J.; Maurer, H.; Reiser, M. Verbal-motor attention-focusing instructions influence kinematics and performance on a golf-putting task. *J. Mot. Behav.* **2014**, *46*, 309–318. [CrossRef]
8. Sim, M.; Kim, J.U. Differences between experts and novices in kinematics and accuracy of golf putting. *Hum. Mov. Sci.* **2010**, *29*, 932–946. [CrossRef]
9. Karlsen, J.; Smith, G.; Nilsson, J. The stroke has only a minor influence on direction consistency in golf putting among elite players. *J. Sports Sci.* **2008**, *26*, 243–250. [CrossRef]
10. Merry, C.; Baker, J.S.; Dutheil, R.; Ugbolue, U.C. Do kinematic study assessments improve accuracy & precision in golf putting? A comparison between elite and amateur golfers: A systematic review and meta-analysis. *Phys. Act. Health* **2022**, *6*, 108–123. [CrossRef]
11. Lee, T.D.; Ishikura, T.; Kegel, S.; Gonzalez, D.; Passmore, S. Head–putter coordination patterns in expert and less skilled golfers. *J. Mot. Behav.* **2008**, *40*, 267–272. [CrossRef]
12. Kooyman, D.J.; James, D.A.; Rowlands, D.D. A feedback system for the motor learning of skills in golf. *Procedia Eng.* **2013**, *60*, 226–231. [CrossRef]
13. Mathers, J.F.; Grealy, M.A. Motor control strategies and the effects of fatigue on golf putting performance. *Front. Psychol.* **2014**, *4*, 1005. [CrossRef]
14. Delay, D.; Nougier, V.; Orliaguet, J.-P.; Coello, Y. Movement control in golf putting. *Hum. Mov. Sci.* **1997**, *16*, 597–619. [CrossRef]
15. Dias, G.; Couceiro, M.; Mendes, P.; Gomes, R.; Mendes, R.; Vaz, V.; Martins, F.; Gama, J.; Castro, M.A. Golf-putting performance in skilled golfers at different distances to the hole. *Appl. Sci.* **2021**, *11*, 11785. [CrossRef]
16. Tanaka, H.; Iwami, M. Estimating putting outcomes in golf: Experts have a better sense of distance. *Percept. Mot. Ski.* **2018**, *125*, 313–328. [CrossRef] [PubMed]
17. Hasegawa, Y.; Fujii, K.; Miura, A.; Yamamoto, Y. Resolution of low-velocity control in golf putting differentiates professionals from amateurs. *J. Sports Sci.* **2017**, *35*, 1239–1246. [CrossRef]
18. Vicon Motion Systems Limited. *Plug-In Gait Reference Guide*; Vicon Motion Systems Limited: Oxford, UK, 2023; p. 110.
19. HAS Motion. Plug-In Gait. Available online: https://www.has-motion.com/download/IORGaitFiles/pigmanualver1.pdf (accessed on 1 November 2024).
20. Brown, S.J.; Selbie, W.S.; Wallace, E.S. The X-Factor: An evaluation of common methods used to analyse major inter-segment kinematics during the golf swing. *J. Sports Sci.* **2013**, *31*, 1156–1163. [CrossRef]
21. Joyce, C.; Burnett, A.; Ball, K. Methodological considerations for the 3D measurement of the X-factor and lower trunk movement in golf. *Sports Biomech.* **2010**, *9*, 206–221. [CrossRef]

22. Wu, G.; van der Helm, F.C.; Veeger, H.E.; Makhsous, M.; Van Roy, P.; Anglin, C.; Nagels, J.; Karduna, A.R.; McQuade, K.; Wang, X.; et al. ISB recommendation on definitions of joint coordinate systems of various joints for the reporting of human joint motion--Part II: Shoulder, elbow, wrist and hand. *J. Biomech.* **2005**, *38*, 981–992. [CrossRef]
23. Bieńkiewicz, M.M.N.; Bringoux, L.; Buloup, F.; Rodger, M.; Craig, C.; Bourdin, C. The limitations of being a copycat: Learning golf putting through auditory and visual guidance. *Front. Psychol.* **2019**, *10*, 92. [CrossRef]
24. Pataky, T.C.; Vanrenterghem, J.; Robinson, M.A. The probability of false positives in zero-dimensional analyses of one-dimensional kinematic, force and EMG trajectories. *J. Biomech.* **2016**, *49*, 1468–1476. [CrossRef]
25. Pataky, T.C. One-dimensional statistical parametric mapping in Python. *Comput. Methods Biomech. Biomed. Eng.* **2012**, *15*, 295–301. [CrossRef]
26. Pataky, T.C. SPM1D. Available online: https://spm1d.org/ (accessed on 2 September 2020).
27. Couceiro, M.; Dias, G.; Araújo, D.; Davids, K.; Gama, J.; Mendes, R.; Martins, F.; Vaz, V. The re-organization of action in golf putting under different task constraints. *J. Mens Health* **2022**, *18*, 026. [CrossRef]
28. Wu, Y.-L.; Huang, C.-F.; Marquardt, C.; Wang, H.-T. Putting performance and kinematics differ with skill level in female golfers. *Open Sports Sci. J.* **2020**, *13*, 20–26. [CrossRef]
29. Healy, A.; Moran, K.; Dickson, J.; Hurley, C.; Smeaton, A.; O'Connor, N.; Kelly, P.; Haahr, M.; Chockalingam, N. Analysis of the 5 iron golf swing when hitting for maximum distance. *J. Sports Sci.* **2011**, *29*, 1079–1088. [CrossRef] [PubMed]

Disclaimer/Publisher's Note: The statements, opinions and data contained in all publications are solely those of the individual author(s) and contributor(s) and not of MDPI and/or the editor(s). MDPI and/or the editor(s) disclaim responsibility for any injury to people or property resulting from any ideas, methods, instructions or products referred to in the content.

www.ingramcontent.com/pod-product-compliance
Lightning Source LLC
Chambersburg PA
CBHW062045220926
51850CB00010B/2669
* 9 7 8 3 7 2 5 8 7 7 4 3 0 *

MDPI AG
Grosspeteranlage 5
4052 Basel
Switzerland
Tel.: +41 61 683 77 34

Bioengineering Editorial Office
E-mail: bioengineering@mdpi.com
www.mdpi.com/journal/bioengineering

Disclaimer/Publisher's Note: The title and front matter of this reprint are at the discretion of the Guest Editors. The publisher is not responsible for their content or any associated concerns. The statements, opinions and data contained in all individual articles are solely those of the individual Editors and contributors and not of MDPI. MDPI disclaims responsibility for any injury to people or property resulting from any ideas, methods, instructions or products referred to in the content.